Breakfast Cereals
And How They Are Made

Edited by
Robert B. Fast and Elwood F. Caldwell

Published by the
American Association of Cereal Chemists, Inc.
St. Paul, Minnesota, USA

Cover photographs: Front, Malt-O-Meal Co., Northfield, MN
Back, Rainbow Foods, Inc., Minneapolis, MN

Library of Congress Catalog Card Number: 89-082452
International Standard Book Number: 0-913250-70-8

Printed in the United States of America

American Association of Cereal Chemists
3340 Pilot Knob Road
St. Paul, Minnesota, 55121 USA

Author-Contributors
and Their Affiliations

Benjamin Borenstein, Hoffmann-LaRoche (*retired*), Teaneck, NJ 07666

Robert E. Burns, Spray Dynamics Division, Par-Way Manufacturing Co., Costa Mesa, CA 92627

Elwood F. Caldwell, American Association of Cereal Chemists, St. Paul, MN 55121

Myrland J. Dahl, Malt-O-Meal Co., Northfield, MN 55057

Robert B. Fast, Robert B. Fast Associates, Inc., Poultney, VT 05764

Susan J. Getgood, Wolverine Corp., Merrimac, MA 01860

Howard T. Gordon, Hoffmann-LaRoche, Inc., Nutley, NJ 07110

Leonard E. Johnson, Hoffmann-LaRoche, Inc., Nutley, NJ 07110

Theodore P. Labuza, Dept. of Food Science and Nutrition, University of Minnesota, St. Paul, MN 55108

Charles Lauhoff, Lauhoff Corp., Detroit, MI 48207-4493

George H. Lauhoff, Lauhoff Corp., Detroit, MI 48207-4493

Haines B. Lockhart, The Quaker Oats Co. (*retired*), Crystal Lake, IL 60014

Britton D. F. Miller, Food Engineering Corp., Minneapolis, MN 55441

Robert C. Miller, Consulting Engineer, Auburn, NY 13021

Edward J. Monahan, General Mills, Inc. (*retired*), Golden Valley, MN 55422

Robert O. Nesheim, Advanced Health Care, Monterey, CA 93940

Scott E. Seibert, National Oats Co., Cedar Rapids, IA 52402

Fred J. Shouldice, Shouldice Bros., Inc., Battle Creek, MI 49015

John E. Stauffer, Stauffer Technology, Greenwich, CT 06830

Donald D. Taylor, Wolverine Corp., Merrimac, MA 01860

William J. Thomson, Spooner Inc., Williamsville, NY 14221

Dedication

When the editors of this book began their research careers in the breakfast cereal industry, they found a near void of literature on the subject of how various kinds of hot and cold cereals were produced. There were a few patents in print and some company literature, but very little else existed on general manufacturing processes.

A great deal of our basic knowledge came as the result of the patience, understanding, encouragement, and willingness to share knowledge on the part of those associates who preceded us—individuals such as Bob Hower, Maurice Bevier, Leonard Foster, Al Greymore, Gurney White, Dick Graham, Jack Gould, Ben Grogg, Elmer Gustavson, Ed Lilly, and Eldor Rupp. It is to those individuals or to their memory that we dedicate this book.

We could name many more, but space does not permit doing this in a brief dedication. It is our desire that in some small way our efforts in bringing this material together will in a like manner assist new entrants to the industry in understanding how breakfast cereals are made.

Preface

The rise of a breakfast cereal industry over the past century and the proliferation of brands and types of cereals over recent decades both owe much to creative advertising and marketing. However, neither would have been possible without accompanying developments in technology, which is what this book is about. Much has been *said* about technological advances in the conversion of bulk commodities like wheat, corn, oats, and other grains into those varied, attractive, delicious, and nutritious packaged products that fill supermarket shelves. However, not much has been *written*, at least in such a way as to bring together in any detail the diverse technologies involved. Although the separate operations are not highly proprietary in themselves, the ways they are combined and the formulations used to yield distinctive products have traditionally been regarded as highly proprietary.

By contrast, details of other cereal grain processes (flour milling, for example) have become well known, widely practiced, much taught, and much written about. Perhaps because of the proprietary aspect, this has essentially never happened in the breakfast cereal area. Short courses, single chapters in monographs or authored books about the technology of the cereal grains, literature from suppliers of equipment items and of specialized ingredients, and word of mouth discussion have served to record and communicate what is generally known about breakfast cereal manufacture.

We hope this book will change all that. It does *not* reveal any closely guarded secrets of present manufacturers, because neither we nor our other author-contributors know what those are in any detail, nor would they be disclosed if we did. But what the book *does* do is provide, for the first time ever in a single volume as far as we know, a comprehensive view of what is in its title—breakfast cereals and how they are made.

In putting a manuscript together over the past three years, we have drawn upon our own experience and that of others to provide an *integrated* treatment of the subject. We gratefully acknowledge the indispensable contributions of the 19 others besides ourselves who have contributed paragraphs, sections, or chapters directly. However, responsibility for the collation and editing of these and of integration of the material into a useful whole is ours, including the blame for any errors or oversimplifications that expert readers may detect. The intent has been to provide an introduction to breakfast cereal technology that will be useful to newcomers with no background in the field and at the same time assist more experienced persons—product development scientists, process engineers, technical salespersons, nutritionists, sensory analysts, packaging technologists, and others—by providing a perspective across their various fields as applied to the development, manufacture, and marketing of breakfast cereals.

Structurally, the book is oriented primarily to the ready-to-eat products that dominate the breakfast cereal market around the world, wherever there is such a market. Thus, it begins by introducing readers briefly to the principal cereal grains, then gives an overview of the technologies involved in turning them into attractive edible products. Next come the equipment and operational details, covered in five chapters on processing and one on packaging. Formulation is discussed only in broad terms—this is not a recipe book. In order to keep it to one volume, we have elected not to deal in any detail with the myriad of colors, flavors, and other comparable ingredients that we might have, leaving that for another book and other authors who might wish to tackle it.

After pausing to summarize processing and packaging information about hot cereals in a single chapter, we then conclude with one chapter each on the important general subjects of fortification and preservation of cereals, cereal nutrition, and quality assurance in cereal plants. Two important appendixes are included—one listing some of the manufacturers of the principal items of cereal processing and packaging equipment and another picking up additional references beyond those attached to each chapter.

We gratefully acknowledge not only our cooperating author-contributors but also those persons who reviewed sections and chapters as they were written, including some who were themselves contributors to other parts of the book. In particular we thank Jeff Culbertson, Kim Doble, Carl Hoseney, Julie Jones, Ted Labuza, Leon Levine, Bob Miller, Richard Sachse, Mike Scott, Scott Seibert, Frank Wilgen, Bill Zuse, and others who reviewed or discussed parts of the manuscript as it developed. We also thank Marie McHenry, secretary, for her patient assistance and concern for accuracy and members of the AACC editorial and production staff for their advice and technical assistance.

Robert B. Fast
Elwood F. Caldwell

Contents

Chapter 1

The Cereal Grains

ELWOOD F. CALDWELL
ROBERT B. FAST

Breakfast cereals have been defined as "processed grains for human consumption" (Fast, 1987). One or more of the cereal grains or milled fractions thereof are indeed major constituents of all breakfast cereals, approaching 100% in the case of cereals for cooking. The proportion drops well below this in many ready-to-eat cereals, and to less than 50% in presweetened products having nutritive sweeteners (sucrose, or fructose, dextrose, and other corn-based products) as substantial constituents.

In broad terms, then, breakfast cereal ingredients may be classified as 1) grains or grain products, 2) sweeteners, caloric or otherwise, 3) other flavoring or texturizing macroingredients, 4) microingredients for flavor and color, and 5) microingredients for nutritional fortification and shelf life preservation. Each of these areas is itself the subject of entire books, and even the cereal grains can be dealt with only briefly here in the specific context of the manufacture of breakfast cereals. Accordingly, this introductory chapter summarizes the characteristics and the role of the principal grains used in breakfast cereals (corn, wheat, oats, rice, and barley). Brief comments are included on the other principal and traditional macroingredients malt, sugar, and salt, the first two of which also are or can be of cereal origin. Discussion of nutritional and preservative microingredients (the latter primarily antioxidants) is reserved for Chapter 10, and discussion of other flavor and color ingredients is reserved for other books by other authors, except insofar as they are dealt with along the way in chapters on other aspects of breakfast cereals.

Corn

Corn (*Zea mays* L.), or maize, as it is known in Europe and other parts of the world, originated in the Western hemisphere. Although it is now grown around the world, production of field corn in the United States exceeds that in any other country and is usually about equal to that of the rest of the world put together. Illinois and Iowa lead all other states, but Nebraska, South Dakota, Minnesota, Indiana, and Ohio also have substantial crops, and some is grown in every state.

Field corn is entirely different as a food crop from the sweet corn that is harvested immature in summer and cooked fresh on the cob or cut from the cob in food plants and preserved by canning or freezing for later consumer use as a vegetable. Field corn is allowed to mature before harvest in the fall. It is shelled as a part of the harvesting operation and typically requires drying on the farm before storage or delivery to country elevator, grain terminal, or mill.

A very small part of the field corn crop is of white corn varieties, used for making the hominy grits and white cornmeal sold at retail in the United States, the grits being for use as a vegetable menu item or hot breakfast cereal, and the meal for making white corn bread and muffins. However, since corn is grown primarily for animal feed, for which the yellow varieties are preferred, availability and economics dictate that yellow corn must be the grain of choice for most food ingredient uses, including breakfast cereals.

As will be pointed out in more detail in other chapters, the characteristic product milled from yellow corn and used for making ready-to-eat corn flakes by the traditional process is known as "flaking grits." The corn is dry-milled to separate the germ and bran from the endosperm, the larger particles of which go into the flaking grits, which are then cooked (with other ingredients), dried, tempered, flaked, and toasted. Progressively smaller endosperm particles form corn meal and corn flour, which can be used as ingredients in extruded corn or mixed-grain cereals. The bran fraction, for many years used only in animal feed, and the germ press-cake, after extraction of its oil content, are now recovered and made available in an edible grade as a fiber ingredient for cereals and other food products.

Details of the milling and other aspects of corn as a grain and a source of food ingredients have been provided by Watson and Ramstad (1987). The corn used for dry milling into grits and for wet milling is typically U.S. No. 1 or No. 2 yellow dent corn. Special varieties of corn are grown and processed in which the starch is all amylopectin (waxy corn) or contains 50 or 70% amylose, instead of the 25:75 amylose-amylopectin ratio of ordinary market corn. These varieties are not known to be used in breakfast cereals, although the waxy type might be expected to give greater expansion in extruded products. The official U.S. grade requirements for corn appear in Table A-1 of Chapter 12. Typical specifications for flaking grits and for corn meal and flour compiled from that table and other sources are given in Table 1.

Wheat

Wheat was an established food crop at the dawn of history. It probably originated from grasses native to the Middle East, but cultivation of bread wheats was common in Europe by the time of the Greek and Roman empires, and wheat was brought to America by the earliest explorers. It is grown in temperate zones around the world and on some part of every continent except Antarctica.

The principal species of wheat are *Triticum vulgare* or bread wheat; *T. durum*, which has extra hard kernels used primarily for macaroni and related pasta products; and *T. compactum* or club wheat, which has very soft kernels. There are numerous varieties and cultivars within each species. Wheat is also classified in the United States according to whether it is hard or soft, white or red, and planted in winter or spring. It is also graded according to such criteria as test weight per bushel and content of damaged kernels, foreign materials, and wheat of other classes. Official U.S. grade requirements for wheat appear in Table A-3 of Chapter 12.

For shredded wheat, the grain of choice is usually No. 2 soft white wheat. Because of the lack of color in the bran, this makes the best-appearing product. White wheats adaptable to the shredding process are grown in several areas, particularly in western New York state.

TABLE 1
Typical Composition (as-is basis)
of Dry-Milled Corn Products[a]

Component	Flaking Grits	Corn Meal	Corn Flour
Moisture, %	11.7	11.5	13.0
Protein, % (N × 6.25)	7.0	7.5	5.2
Fat, %	0.6	0.7	2.0
Crude fiber, %	0.2	0.2	0.5
Ash, %	0.2	0.3	0.4
Starch, %	78.3	78.0	76.4
Other polysaccharides, %	2.0	1.8	2.5
Particle size			
Maximum, through U.S. No.	3.5	30	60
Minimum, over U.S. No.	6	60	325

[a]Adapted from Alexander (1987).

For flaking, soft red wheat kernels are satisfactory. The added color in the bran is not a detriment in wheat flakes, since the malt and sugars added for flavor development already result in a brown color that is considered desirable.

Both red and white wheat brans are used for bran flakes and other bran cereals. These are by-products of the flour milling industry and perform equally well in most processes, the choice being a matter of supply, price, and the exact appearance of the finished product that is desired.

Durum wheat is best for puffing. The kernels are typically large and make large puffs. The bran is light colored and results in a more appetizing puffed appearance when it is from durum rather than from hard red or red durum wheats. Even so, for the best appearance, wheat for puffing should be pearled to remove a part of the bran, as it shatters rather than expands when the wheat is puffed.

Hard red wheats are used for some hot whole wheat cereals because they tend to retain their granular character during home preparation, which results in a less pasty and more interesting texture and mouthfeel.

Hard red wheats are also usually the basis for farina, or more accurately, for the first and second middlings product obtained from the flour milling industry to be marketed as farina or as one of the proprietary flavored farina hot-cereal products that are marketed to consumers. However, since hard red wheats are grown primarily for the gluten strength and baking quality of the flours made from them, the wheats themselves and the milling fractions from them may command a premium price over other wheats.

Details of the milling and other aspects of wheat as a grain and a source of food ingredients have been provided by Pomeranz (1988). A typical specification for farina is given in Chapter 9.

Oats, Rice, and Barley Compared with Corn and Wheat

The processing of oats, rice, and barley contrasts with that of corn and wheat in that they are shipped from the farm "covered," i.e., with the hull intact, necessitating removal of the hull as the initial milling operation, and in that separation of their kernel components (into bran, endosperm, and germ) is relatively difficult and done only minimally in milling. The processing of oats, rice, and barley has been compared briefly by Hoseney (1986).

Oats and barley are similar to corn in that much of the crop goes to animal feed, whereas rice is similar to wheat in being grown primarily for human consumption. Within uses for human consumption, corn, rice, and barley all find significant use as fermentation substrates for alcoholic beverages. As pointed out earlier, corn also is the basis of a wet-milling industry, the products of which include starch and a variety of carbohydrate sweeteners, and extraction of its germ provides one of our major edible oils.

These usage differences are more the result of growing patterns and of morphology and processing differences than the result of differences in overall chemical composition, as indicated by the approximate comparative data given in Table 2.

Oats

Oats are known to have been cultivated since about 2,000 B.C. Over 75% of the world crop consists of cultivars of *Avena sativa* ("white oats") or *A. byzantina* ("red oats"), which are either spring or winter crops depending on the location and climate. The largest share of the world crop is grown in the USSR, followed traditionally by the United States, Germany, Poland, and Canada. However, the U.S. crop declined precipitously in amount during the 1980s due to price supports and acreage allotments favoring other grains.

In the United States, a majority of the planted oat crop (about 60%) never leaves the farm, being harvested for feed use or serving as a forage crop. The remaining 40% that is harvested and shipped in commerce is believed to go about equally to milling for human food and to the racehorse trade. Most of the oats grown in the United States come from the upper Midwest, specifically North and South Dakota, Minnesota, and Wisconsin. These four states represented 63% of the total U.S. production in the 1980–1988 period. All of the major U.S. mills processing oats for human consumption are located in or adjacent to these states.

TABLE 2
Comparison of Typical Composition of Cereal Grains[a]

Component	Corn	Wheat	Oat Groats	Brown Rice	Pearled Barley
Moisture, %	13.0	12.5	8.3	12.0	11.1
Protein, %	8.9	12.3	14.2	7.5	8.2
Fat, %	3.9	1.8	7.4	1.9	1.0
Crude fiber, %	2.0	2.3	1.2	0.9	3.2
Ash, %	1.2	1.7	1.9	1.2	2.5
Thiamin, mg %	0.37	0.52	0.6	0.34	0.12
Riboflavin, mg %	0.12	0.12	0.14	0.05	0.05
Niacin, mg %	2.2	4.3	1.0	4.7	3.1

[a]Adapted from Barr et al (1955).

In purchasing oats, high product quality and milling yield are the most important considerations, in that order. Of the best oats, 160 lb is required to yield 100 lb of hulled kernels or groats, but up to 200 lb of normal oats may be required. Official U.S. grade requirements for oats appear in Table A-2 of Chapter 12. However, the best oats for milling exceed the minimums for Grade 1 in certain respects, having a test weight of 38, a minimum sound count of 96%, and a maximum of 3% foreign material (including a maximum of 1.5% of other grains, with barley being especially objectionable).

The protein level and protein efficiency ratio of oats are among the highest of the cereal grains. Although these are positive attributes in terms of human nutrition, they do not enter into the grading standards, nor are they pricing considerations except by special contract. This reflects the traditional role of oats as a source of calories rather than of protein for animal feed, with crop breeders and growers interested in yield and grade but not protein. As a result, the U.S. crop is lower in protein by 2% or more than it was earlier in the century, averaging 15.2% for the crop years 1986–1988. Oats from other countries are even lower; those from Canada average 13.6% and those from Argentina 12.1% for the same period.

Similar variations in other attributes such as dietary fiber and β-glucan have been reported informally. Accordingly, to the extent that such analytical parameters might be important to cereal processors, advance agreement on them or careful examination of representative samples could be important purchase considerations, in addition to the traditional grade standards.

All aspects of the production, composition, and processing of oats are covered by Webster (1986).

Rice

Rice, as *Oryzae sativa* L., is a leading food crop, the staple food of over half the world's population. Although it is generally considered (and grown as) a semiaquatic annual grain plant, varieties grow in areas from deeply flooded land to dry, hilly slopes, and (like wheat) on every continent except Antarctica. The gross yield of rice per unit of land area is second only to that of corn, but its net food yield is the highest of any cereal grain.

The harvested grain before hulling is known as rough rice. The world crop for 1988–89 was estimated to be 477 million metric tons, with China and Taiwan in the lead with 36.7% of this, followed by India (19.8%), Indonesia (9.0%), Thailand (4.2%), Vietnam (3.1%), Japan and Burma (2.6% each), Brazil (2.3%), Philippines (1.9%), South Korea (1.8%), and the United States (1.5%). The remaining 14.4% was grown in other areas, including the USSR, eastern Europe, and Africa, which may also be net importers. Rice is overwhelmingly consumed where it is grown. Despite their modest share of world production, Thailand and the United States are the principal exporters of rice, with their shares at 38 and 18%, respectively, followed by China with 10% of world exports.

Like corn starch, rice starch has amylose and amylopectin components. In the United States, waxy (low-amylose) short- and medium-grain rices are grown primarily in California. The medium-grain varieties are useful for puffing to make a breakfast cereal, as well as for brewing adjuncts and for parboiling. Medium-grain or long-grain varieties with intermediate levels of amylose are grown primarily in Texas, Louisiana, and Arkansas. These provide milled rice for breakfast cereals, for canning, and for direct consumer home use. Some of the milled rice is also further processed to make quick-cooking or instant rice.

"Wild rice" is not really rice at all, but a different aquatic plant, *Zizania palustris*, whose seeds resemble rice except in color, but whose growth characteristics are otherwise unlike those of any of the cereal grains.

MILLING

Harvested at 18–25% moisture, rice is mechanically dried to under 15% moisture on the farm, at a commercial dryer, or at the mill before storage or processing. Rough cleaning (removal of sticks, stones, dust, and other foreign material) may precede drying, which is usually done in upright continuous flow dryers with a concurrent flow of heated air. After cooling, the dried rice is cleaned by various separation methods to give rough rice or "paddy."

Some paddy is parboiled or "converted" by being soaked in water, then drained, steamed to gelatinize the starch, and dried again. This traditional process may increase the nutritive value by distributing soluble B vitamins from the bran into the endosperm, and it may stabilize against insects and lipolytic rancidity, but it does not decrease consumer cooking time.

Most paddy is hulled between rubber rolls turning in opposite directions, aspirated to remove hulls, and separated from the unhulled or rough rice in a paddy machine or gravity table to give brown rice. Most parboiled rice is prepared from brown rice by the steps already outlined for parboiling.

Finally comes the most important rice milling operation, removal of some or all of the bran layers by abrasion, or "pearling," as it is called. After removal of the pearlings by aspiration, the milled or white rice is separated by size into heads (unbroken kernels), second heads (larger broken kernels), brewers' rice (smallest broken kernels), and screenings. Although these grades have other even more important uses, head rice is used for puffing and second heads for rice flakes. If rice is desired as a component of a formulated flaked, shredded, or extruded cereal, rice flour made from grinding second heads or brewers' rice is used.

Official U.S. grade requirements for rice appear in Table A-4 of Chapter 12. However, these are little used domestically by U.S. rice millers, who prefer to market based on their own grade standards. The official grades can be and are used in contracting for export orders.

Details of the production, processing, nutritional, and application aspects of rice as a grain and source of food may be found in Juliano (1985), where particular emphasis is given to international aspects.

Barley

Like wheat, barley was already an established food crop when our earliest historical records originated, and it may have been cultivated as long ago as 15,000 B.C. It grows over a broader environmental range than any other cereal grain. In addition to this broad ecological adaptation, other distinctive characteristics include almost equal utility as food and feed and superior adaptability to malting for brewing purposes.

As a world crop, its production is somewhat less than half that of the other principal cereal grains (42% of corn, 38% of wheat, 43% of rice). This consists primarily of subspecies of *Hordeum vulgare* L.; these may be two-rowed or six-rowed types according to the seed distribution on the plant spikelets.

Over 30% of the world crop is grown in the USSR, followed by France, Canada, the United Kingdom, the United States, and Germany with 5–7% each. Within the United States, five states (North Dakota, Idaho, Montana, California, and Minnesota) produce over two thirds of the crop; all of these use spring planting. The grain is split about equally between animal feed and malting, with only about 2% for other food uses and up to 5% for seed. As with oats, much of the feed use is on the farm where grown; thus most of the barley sold commercially is for malting. This has significance for breakfast cereals because of the use of malt flours and extracts as flavoring materials.

MALTING

Processing steps in the conversion of barley to malt and the biochemistry involved were described in some detail by Pomeranz (1975) in a review that is still applicable although published some years ago, and they also have been summarized by Hoseney (1986). The fundamental purpose of malting is to allow the development of the α-amylase activity that accompanies the sprouting of all cereal grains. When enzyme activity is at a maximum, the sprouted grains are kilned and dried so as to stop further growth and render the sprouts friable without destroying the enzyme activity developed in the kernel. The dried kernels after removal of the sprouts form the malt. In addition to enzyme activity, it has a characteristic flavor developed during kilning.

BREWING

For brewing, the whole malt is ground and cooked with water and frequently other cereal adjuncts in a process known as mashing. The products of this are wort, the distinctively flavored dilute extract in which the complex carbohydrates of the cereal mash have been enzymatically converted to fermentable sugars, and the spent grains. After separation from the spent grains, the wort is boiled to sterilize it; hops and yeast are added; and the mixture is fermented to make beer.

FLAVORING WITH MALT

Apart from its function in brewing beer, malt is a desirable flavor in other food products, with or without the enzyme activity. Barley malt can be milled to separate the hulls and provide a meal or flour product for use in dry formulations, or it can be extracted with water (as in mashing) and the resulting extract concentrated to a syrup or dehydrated to a powder for use in flavor formulations. In formulating with any of these preparations, care must be taken to specify whether the desired product is one that retains its enzyme activity—i.e., is "diastatic"–or does not retain enzyme activity ("nondiastatic"). Diastase was the original name for malt amylase, in use before it was known to consist of both α- and β-amylases. Nondiastatic products are usually used for flavoring breakfast cereals, as otherwise an unwelcome liquefaction of the starchy components of a formula may occur.

Malt syrup is normally a brown viscous liquid with poor flow properties that is typically diluted with water before being added to the grain. It is available in degrees of color and flavor concentration and results in an attractive light brown color developing in cereals during cooking. However, the use of too much in a formula may result in bitterness in the finished product.

MILLING OF BARLEY

The milling of barley for food uses other than malting resembles that of oats more than of wheat or corn in that it consists of cleaning and sizing, hulling (for barley, by pearling or a scouring type of decortication), sieving and aspiration (to remove bran and make pot or pearl barley, used for thickening soups), and flaking or grinding (to make flakes, grits, or flour from pearl barley). Barley flakes or grits are a key ingredient in some baked goods and cereals, and barley flour is used in the formulation of some infant cereal foods. Chemical and screen analyses of typical medium barley grits are given in Table 3.

Other Cereal Grains

Little or no rye or grain sorghum is used in formulating breakfast cereals. Rye may be used when mixed with other grains. Milling is similar to that of wheat, but all rolls are corrugated and purifiers are not used. Generally two types of rye flour are produced, light (80% of the total) and dark (the remaining 20%). The production, chemistry, and technology of rye have been reviewed by Bushuk (1976).

TABLE 3
Chemical and Screen Analyses
of Typical Medium Barley Grits

Component	As Specified	Typical Value
Moisture, %	13.0 max.	11.5
Protein, % (N × 6.25)	9.0 min.	10.5
Crude fiber, %	1.5 max.	0.7
Fat, %	1.0 min.	1.2
Ash, %	1.0–2.0	1.3
Screen (on Ro-Tap, 3 min)		
On U.S. No. 8, %	2.0 max.	0.2
On U.S. No. 12, %	35.0 ± 15	37.0
On U.S. No. 16, %	45.0 ± 20	45.0
On U.S. No. 20, %	15.0 ± 10	17.0
Through U.S. No. 20, %	3.0 max.	0.8

Other Traditional Ingredients
of Breakfast Cereals

SWEETENERS

In a past era, the term "breakfast cereal" connoted hot cooked products as much as it did corn flakes, wheat flakes, or any of the other now-common ready-to-eat cereals. The hot cereals of the time, primarily rolled oats and farina, never reached the table (and frequently never the home) in packaged form. They appeared in branded, consumer-packaged form only about the turn of the century and continued to be available in bulk for many years after that. If a cereal package was on the table, it was probably corn flakes, with an "ingredient list" consisting of the statement "Flavored with malt, sugar, and salt."

Although even corn flakes now has a more complex listing, especially when nutritionally enriching or fortifying ingredients are included, the flavor and color basis of many ready-to-eat breakfast cereals can still be summarized as "malt, sugar, and salt." The role of malt was discussed in the section of this chapter on barley. We next turn to the various sweeteners now fulfilling the role of sugar in the malt, sugar, and salt trio.

Sucrose

Walker (1989) has reviewed the use of sucrose in ready-to-eat breakfast cereals, concluding that sugars are an essential functional ingredient in most of them, forming a substantial part of the composition of presweetened cereals as well as influencing their sensory character. Sucrose is frequently preferred for surface applications because it can either crystallize as a frosty surface or form a hard glaze, as desired. Such surface applications represent the major use of sucrose in cereals, although smaller amounts are found in the base formula of most cereals not regarded as presweetened (shredded wheat, puffed wheat, and puffed rice being notable exceptions).

Sucrose can be from either sugar cane or sugar beets. A fine granulated form may be needed for certain purposes, but most large operations prefer to use sucrose as a 67° Brix liquid.

Brown sugars may be used for flavor. They are available in a range of color and flavor intensities, since most cane or beet sugar refiners produce them by reincorporating cane sugar molasses into purified white sugar. They may cake during storage and require milling or regrinding before use.

Invert syrup (cane or beet sugar hydrolyzed to its monosaccharide components glucose and fructose) may be used in a formulated cereal flake, shred, or extrudate to cause more browning during cooking and toasting. Pound for pound, it is also sweeter than sucrose, since the greater sweetness of fructose relative to sucrose outweighs the lesser sweetness of glucose relative to sucrose.

Corn Sweeteners

Corn sweeteners also can be either dry or liquid. The functional properties of maltodextrins, corn syrup solids, and corn syrup have been reviewed by Luallen (1988), along with a discussion of native and modified starches.

Maltodextrins and corn syrup products are identified primarily by dextrose equivalent (DE), which gives their percentage reducing power calculated against that of pure glucose or dextrose. Maltodextrins run up to 15 DE, and corn syrup solids typically run from 20 to 42 DE. Conventional liquid corn syrups run from 42 to 65 DE, with the "regular" or low-conversion product being 42 DE. As with invert syrup, such products used in the base formula of a cereal cause a browning reaction during cooking and toasting. Corn syrup is also used in coating formulas to control sucrose crystal size.

The use of fructose in breakfast cereals has been reviewed by White and Parke (1989). Fructose was at one time available in quantity only as a component of invert syrup, but the 1980s saw the introduction on a commercial scale of high-fructose corn syrup (HFCS), made by enzymatic isomerization of the dextrose in corn syrup to fructose. HFCS can be made as high as 90 DE. According to White and Parke (1989), HFCS at this level has a sweetness of 100% that of sucrose, if 10% solutions are compared at room temperature.

Fructose is also available as a crystalline solid that exhibits sweetness synergy with other sweeteners such as sucrose or aspartame, making possible both a cost and a caloric savings.

Noncaloric Sweeteners

The use of non- or low-caloric sweeteners in cereals does not justify a direct low-calorie claim because their addition does not change the caloric value of the base product. In any event, because of its heat sensitivity, aspartame cannot be used in a formulation to be extruded or toasted; if used, it must be added after heat-processing is complete. The use of saccharin is handicapped by a bitter aftertaste in higher-equivalent-sweetness concentrations as well as a cloudy legal future. Alitame-K is a noncaloric sweetener first approved for use in the United

States in early 1989 although it was already in use in some other countries. It (as well as other products that might be approved in 1990 or later) is said to withstand baking temperatures and thus might see use in cereal formulas if approved for that purpose.

SALT

Of the malt, sugar, and salt trio, salt is usually the lowest in concentration and in cost, but it has a key flavor-blending role—and (in recent and current times) a high and largely unwarrantedly negative nutritional image. In a compilation of certain nutritional labeling values for 79 ready-to-eat cereals in two major supermarkets (Toma and Curtis, 1989), the commonest single declared sodium value was 200 mg/oz, with much clustering in the range of 160–230 mg/oz, suggesting an added salt range of 1.25–1.75%, with some sodium values near zero (e.g., for shredded wheat) and some around 300 mg/oz.

Sodium has been linked for years to hypertension. However, research results quoted by Weaver (1988) indicate that only 15% of the adult U.S. population is sodium sensitive, and other studies have estimated the percentage to be even lower. Breakfast cereals in the amounts normally consumed daily probably affect hypertension minimally if at all, especially in view of the calcium-containing milk that usually accompanies them. Calcium is known to have a hypotensive effect (Weaver, 1988). However, it is also true that cereals are not needed as a source of either sodium or chloride, and it would be a prudent move for the breakfast cereal industry to keep salt content as low as reasonable sensory acceptance would permit.

References

Alexander, R. J. 1987. Corn dry milling: Processes, products, and applications. Page 351 in: Corn: Chemistry and Technology. S. A. Watson and P. E. Ramstad, eds. Am. Assoc. Cereal Chem., St. Paul, MN.

Barr, L., Brockington, S. F., Budde, E. F., Bunting, W. R., Carroll, R. W., Gould, M. R., Grogg, B., Hensley, G. W., Rupp, E. G., Stout, P. R., and Western, D. E. 1955. Facts on Oats. The Quaker Oats Co., Chicago, IL.

Bushuk, W., ed. 1976. Rye: Production, Chemistry, and Technology. Am. Assoc. Cereal Chem., St. Paul, MN.

Fast, R. B. 1987. Breakfast cereals: Processed grains for human consumption. Cereal Foods World 32:241.

Hoseney, R. C. 1986. Principles of Cereal Science and Technology. Am. Assoc. Cereal Chem., St. Paul, MN.

Juliano, B. O., ed. 1985. Rice: Chemistry and Technology, 2nd ed. Am. Assoc. Cereal Chem. St. Paul, MN.

Luallen, T. E. 1988. Structure, characteristics, and uses of some typical carbohydrate food ingredients. Cereal Foods World 33:924.

Pomeranz, Y. 1975. From barley to beer. Bull. Assoc. Oper. Millers. pp. 3503-3512.

Pomeranz, Y., ed. 1988. Wheat: Chemistry and Technology, 3rd ed. Vols. I and II. Am. Assoc. Cereal Chem., St. Paul, MN.

Toma, R. B., and Curtis, D. J. 1989. Ready-to-eat cereals: Role in a balanced diet. Cereal Foods World 34:387.

Walker, C. E. 1989. The use of sucrose in ready-to-eat breakfast cereals. Cereal Foods World 34:398.

Watson, S. A., and Ramstad, P. E., eds. 1987. Corn: Chemistry and Technology. Am. Assoc. Cereal Chem., St. Paul, MN.

Weaver, C. M. 1988. Calcium and hypertension. Cereal Foods World 33:792.

Webster, F. H. 1986. Oats: Chemistry and Technology. Am. Assoc. Cereal Chem., St. Paul, MN.

White, J. S., and Parke, D. W. 1989. Fructose adds variety to breakfast. Cereal Foods World 34:392.

Chapter 2

Manufacturing Technology of Ready-to-Eat Cereals

ROBERT B. FAST

Ready-to-eat (RTE) cereals are processed grain formulations suitable for human consumption without further cooking in the home. They are relatively shelf-stable, lightweight, and convenient to ship and store. They are made primarily from corn, wheat, oats, or rice, in about that order of the quantities produced, usually with added flavor and fortifying ingredients.

Hot cereals, on the other hand, are made primarily from oats or wheat; those made from corn or rice are of minor importance, being produced in relatively small quantities. The original hot cereals required cooking in the home before they were ready for consumption, but now some varieties are preprocessed so that they are ready for consumption with the addition of either hot water or milk to the cereal in the bowl.

The processing of RTE cereals typically involves first cooking the grain with flavor materials and sweeteners. Sometimes the more heat-stable nutritional fortifying agents are added before cooking. Two general cooking methods are employed in the industry today—direct steam injection into the grain mass in rotating batch vessels, and continuous extrusion cooking, with the latter very much on the increase in recent decades. Both of these cooking operations and the equipment commonly used are discussed in detail in Chapter 3.

Most RTE cereals may be grouped into eight general categories for discussion of their manufacturing processes: 1) flaked cereals (cornflakes, wheat flakes, and rice flakes), including extruded flakes, 2) gun-puffed whole grains, 3) extruded gun-puffed cereals, 4) shredded whole grains, 5) extruded and other shredded cereals, 6) oven-puffed cereals, 7) granola cereals, and 8) extruded expanded cereals.

Flaked Cereals

Flaked cereals include those made directly from whole grain kernels or parts of kernels of corn, wheat, or rice and also extruded formulated flakes. The basic objective in making a flaked cereal is to first process the grain in such a way as to obtain particles that form one flake each. We know of no process by which flaked cereals can be made from a large, thin sheet that is broken down to individual flakes after toasting.

Grain selection is therefore very important to the finished character of flaked cereals, and one or more intermediate size reductions and sizing or screening operations may be necessary to provide flakable-sized particles, known as *flaking grits*.

Cornflakes and wheat flakes are typically made from whole-grain kernels or parts of kernels. This practice developed in the early years of the industry, since these materials after processing are the correct size for flaking grits.

With the advent of cooking extruders, however, finer materials, such as flours, can be used, since the flakable grit size is attained by mechanical means. By cooking a dough and forming from it the grits for flaking, much of the equipment for preflaking size reduction and screening in the traditional process can be eliminated.

CORNFLAKES

The best example of a cereal made from parts of whole grain is traditional cornflakes. New varieties of cornflakes made by continuous extrusion cooking have been developed and improved in the 1980s. However, they have not made great inroads into the demand for traditional cornflakes, which have been on the market for about a century. We cover extruded cornflakes later in this chapter.

Formulation

The basic raw material for the traditional cornflake is derived from the dry milling of regular field corn. Dry milling removes the germ and the bran from the kernel, and essentially what is left is chunks of endosperm. The size needed for cornflakes is one half to one third that of the whole kernel. These pieces are raw, unflavored, and totally unsuitable for flaking until they have been processed. They never lose their identity while being converted into flakes. Each finished flake typically represents one grit, although sometimes two small grits stick together and wind up as one flake.

A typical formula for cornflakes is as follows: corn grits, 100 lb; granulated sugar, 6 lb; malt syrup, 2 lb; salt, 2 lb; and water sufficient to yield cooked grits with a moisture content of not more than 32% after allowing for steam condensate. Liquid sucrose at 67°Brix can be substituted for the sugar, with a decrease in the amount of water. Likewise, 26% saturated brine can be used rather than dry salt; however, this solution is very corrosive on pumps and meters. Malt syrup is a very viscous material, and some manufacturers prefer malt flour. The traditional malt syrup is one that does not have any diastatic enzyme activity. Both diastatic and nondiastatic malt flours have also been used.

Mixing

The first step in converting raw corn grits into cornflakes is to mix them with a flavor solution. Master batches of the flavor materials (sugar, malt, salt, and water) may be made up for multiple cooking batches. When this is done, it has been found better to weigh out the correct proportion of flavor syrup to be added to each cooker batch rather than draw it off volumetrically. Temperature and viscosity variations in master batches can result in inconsistent addition of the flavor solution. Every effort should be made to weigh each and every ingredient accurately. Inaccurate and haphazard proportioning of ingredients results in differences in the handling and the quality of grits in subsequent processing steps and ultimately in the quality of the finished product.

Cooking

The weighed amounts of raw corn grits and flavor syrup are charged into batch cookers, which are usually vessels about 4 ft in diameter and 8 ft long. They are capable of being rotated and are built to withstand direct steam injection under pressure. One brand that has been popular is the Johnson cooker, formerly manufactured by the Adolph Johnson Co., of Battle Creek, Michigan. In the United Kingdom, similar pieces of equipment are called Dalton cookers. The Lauhoff Corporation, in Detroit, Michigan, produces a redesigned version of the Johnson cooker. Batch cookers are discussed in more detail in Chapter 3.

The grits and flavor syrup may be loaded simultaneously, or the grits may be added first, the cooker lid closed (or *capped*, as it's called), the rotation started, and then the flavor syrup added. No hard-and-fast rules exist, except that the end result must be a uniform dispersion of flavor throughout the grain mass.

Normally the raw ingredients when fully loaded into the cooker occupy not more than one half to two thirds of its volume, to leave room for expansion during cooking. In cooking corn for cornflakes, it has been found best to increase the batch size so that at the end of the cooking time the cooker is filled to capacity. This batch size, which is slightly larger than that normally used for wheat or rice, produces cooks that are less sticky and easier to process further.

With the grits and flavor in the cooker and the cooker tightly capped, the steam is turned on. The steam quality should be that permitted for food contact. The mass of grits and flavor is normally cooked at 15-18 psi for 2 hr. Some batches take more time than others, as a result of variations in the cooking behavior of the corn grits.

The rotation speed of these batch cookers is usually 1-4 rpm, with the higher rate used for initial mixing only. Too high a speed can lead to attrition of the particles, resulting in slime or mushiness in the cooked product. On the other hand, too low a speed can lead to uneven cooking within a batch.

The moisture content of the cooked mass at the end of the cooking cycle should be not more than 32%. Some batches can be considered well cooked and in good processing condition with a moisture content as low as 28%.

The cooking is complete when each kernel or kernel part has been changed from a hard, chalky white to a light, golden brown and is soft and translucent. A batch is undercooked if large numbers of grain particles have chalky white centers, and it is overcooked if the particles are excessively soft, mushy, and sticky. Properly cooked particles are rubbery but firm and resilient under finger pressure, and they contain no raw starch. Raw starch present after cooking remains through further processing and shows up as white spots in the finished flakes.

When the cooking time cycle is completed, the steam is turned off, and the vent opened to help reduce the pressure inside the cooker back to the ambient pressure and cool its contents. The exhaust may be connected to a vacuum system for more rapid cooling. The cooker is carefully uncapped and the rotation restarted.

Dumping

The cooked food is dumped on a moving conveyor belt under the cooker discharge. Dumping creates an interesting processing problem—that of placing a properly cooked batch of grain, which is optimum at time zero, into the slower continuous flow in the next steps of the process. A batch of cooked corn grits can be dumped from a cooker in about 7 min, but no dryers in the industry can dry them to flakable moisture in 7 min. Almost all processors therefore face the problem of how to get the cooker empty, cool the cooked material to stop the cooking action, and space that material out in a uniform flow to feed a dryer and cooler of reasonable size. While this is being done, the cooker may be needed for the next batch, with its own loading time, steam come-up time, 2-hr cooking time, steam exhaust time, and dump time.

The most common method of solving the dumping problem is to spread the cooked food out over a large area. Some spread it on wide, slow-speed conveyors under the cookers. Others spread it over large areas of perforated plates, with air blowing up through the perforations; these can be stacked to save space and are sometimes agitated.

Delumping

Once on a moving belt, before they are conveyed to a dryer, the cooked grits pass through delumping equipment to break the loosely held-together grits into mostly single grit particles. Delumping is essential to obtain particles or agglomerates of grits small enough for good circulation of heated air around each particle for uniform drying. It may be necessary to accomplish delumping and cooling in steps to get good separation of the grits so that they are the optimum size for drying. In most cases cooling takes place first, to stop the cooking action and remove stickiness from the grit surface. Cooling is kept to a minimum, because in subsequent drying the product is reheated to remove moisture. Most cooling-delumping systems include screening devices. The most common are flatbed gyrating sifters or rotating-wire or perforated-drum screeners.

Drying

From the cooling-sizing operation, the grits are metered in a uniform flow to the dryer. The most prevalent dryer configuration is that of wide, perforated conveyor units passing through a surrounding chamber in which the temperature, humidity, and airflow can be controlled. Dryers and drying in general are discussed in greater detail in Chapter 4.

Drying corn grits is best done at temperatures below 250°F (121°C) and under controlled humidity. It should result in a minimum of skinning over of the particle surface, as this impedes the removal of moisture from the center of the grit. Controlled humidity prevents such case hardening of the grit surface and greatly decreases the time needed for drying to the desired end moisture, usually 10–14%.

Cooling and Tempering

After drying, the grits are put through a cooler to bring them back down to the ambient temperature. Such cooling is usually done in an unheated section of the dryer itself. In certain hot climates or under certain plant conditions some refrigerated air may be required. If the grits are not properly cooled, they darken and lose quality during the next step in the process.

Tempering is merely holding the grits in large accumulating bins to allow the moisture content to equilibrate between grit particles as well as from the center to the surface of individual particles. Other physical and chemical changes also take place within the grit components and affect the degree of blister development in the toasting operation.

In earlier days of processing grits for cornflakes, before controlled-humidity dryers were available, tempering times were long—as long as 24 hr—to allow complete equilibration of moisture. Now, with controlled-humidity drying, tempering times have been reduced to a matter of hours. The moisture content at the end of tempering should be 10–14%.

Flaking

After tempering, the grits are rolled into thin flakes by passing between pairs of very large metal rolls. Flaking rolls are usually made of chilled iron. They are hollow, and the hollow center is fitted with a means of injecting cooling water and removing it once it has served its purpose and taken up heat generated by flaking. Usually, chilled water at a controlled temperature is injected into the shaft at one end of the roll, and the used, heated water is removed from the shaft at the other end. The flow of water through the roll is accomplished by a continuous spiral groove channeled in the interior surface of the roll from one end to the other. Flaking rolls are discussed in detail in Chapter 5.

Tremendous pressures are necessary to flatten the grits into flakes. For normal-sized rolls, 20 in. in diameter and 30 in. long, these pressures are on the order of 40 tons (Matz, 1959).

For normal flaking of corn grits a good temperature at the roll surface is 110–115°F (43–46°C), as measured after the rolls have been used for about 1 hr and are evenly warmed up. Temperatures much over 120°F (49°C) cause excessive roll wear and sticking of the product to the roll surface. Cooling the rolls to temperatures much below 110°F is not necessary, as it does not prolong roll life to any great extent and requires excessive amounts of chilled water. Colder roll surfaces also lack the "grabbing" ability needed to draw the grits into the roll nip, or area where the rolls are closest together.

The moisture content at the time of flaking is most important and has a great bearing on the blister formation, or *development*, of the finished flake. The moisture and the matching oven temperature profile are the two main determining factors in good development, which is generally best in the 10–14% moisture range. To flake cooked corn grits at this low moisture content, it is necessary to steam them or otherwise heat them just sufficiently to make their surfaces sticky enough to allow the rolls to grab them and draw them in.

Roll knives are used to scrape the flakes off the rolls. They must be kept sharp and properly mounted. Roll knives are described more fully in Chapter 5, which also explains roll-feeding devices. These are necessary to maintain a uniform feed of grits, evenly spaced across the whole width of the rolls.

Older-style rolls with babbitt bearings roll on the order of 150 lb of flakes per hour. This tonnage was well matched to the capacity of old-style, direct-fired rotary toasting ovens. Modern rolls, with roller bearings in place of babbitt bearings, can roll between 400 and 800 lb of flakes per hour; the flakes are then toasted in much-upgraded, indirect-fired ovens.

Flaking rolls are mounted directly over the feed end of the oven, usually on the floor or a mezzanine above. It is not unusual to have two pairs of rolls (two-roll stands) feeding one oven.

Toasting

Flakes are usually toasted by keeping them suspended in a hot air-stream rather than by laying them out on a flat baking surface like those used for cookies and crackers. The classical flake-toasting oven is a rotating perforated drum, 3–4 ft in diameter and 14–20 ft long, mounted in an insulated housing. Once matched to the product type and the production rate, the speed of rotation of the drum is rarely varied.

The oven slopes from the feed end to the discharge end. It is important for the slope and the speed of rotation of the drum to be adjusted so that the flakes remain suspended in the air as much as possible and are not thrown out so that they stick to the inside of the drum. The perforations should be as large as possible for good airflow but small enough that flakes do not catch in them and remain there until they burn.

Properly toasted flakes have the correct and desired color and moisture content. Color can be checked visually by the oven operator, and excellent color-measuring meters, which yield numerical results, are also available. A standard acceptable range from too light to too dark can easily be established with these instruments. They may be read and the readings manually recorded on quality control charts on a regular basis, and closed-loop automatic recording and control are also possible with state-of-the-art instrumentation.

The moisture content of flakes is usually in the range of 1.5–3%. Checking is done both by feel and by moisture meter. Like color-measuring meters, moisture meters can be read and the readings recorded on quality control charts, or they can be linked to automatic controls. Properly trained oven operators can become very expert at judging product quality, but the use of meters is strongly recommended for constant quality. For properly toasted cornflakes, oven temperatures in the range of 525–625°F (274–329°C) are usually employed, and the residence time is about 90 sec.

Another style of toasting oven also suspends the flakes in heated air, but they are carried through the oven by a vibratory trough from the feed end to the discharge. These ovens can be zoned or coupled together to provide varying oven temperature profiles for varying toasting effects. Oven styles are covered in greater detail in Chapter 5, and nutritional fortification in Chapters 7 and 10.

WHEAT FLAKES

The processing of wheat flakes is different from that of cornflakes because of differences between the grains. For cornflakes, the starting material is broken chunks of corn kernel endosperm from which the bran has been removed, but for wheat flakes, the starting material is whole wheat kernels with all seed parts intact (germ, bran, and endosperm).

Preprocessing

The object of the cooking process is the same for wheat flakes as for cornflakes: complete gelatinization of all the starch present and even distribution of the flavors (sugar, salt, and malt) throughout the individual kernels. This requires a preprocessing step before cooking, since the bran coat is a barrier that keeps the water and flavor materials from penetrating to the interior of the kernel.

The kernels are broken open by a process referred to as *bumping*; the grain is lightly steamed and then run through a pair of rolls to crush the kernels slightly. If the crushing and flattening are too severe, the resulting flour and fine material cause the cook to be excessively soft and gluey and difficult to process further.

Formulation

Wheat used for the production of wheat flakes is usually soft red or white winter wheat. Rarely is it necessary to pay the premium prices for hard wheat for wheat flake production.

A typical formula for wheat flakes is as follows: bumped wheat, 100 lb; fine-granulated sugar, 8–12 lb; malt syrup, 2 lb; salt, 2 lb; and water sufficient to yield a cooked product with a moisture content of 28–30%, including steam condensate. The flavor materials and water can be made up in a master batch sufficient for multiple cooks, and aliquots then are weighed for each cook.

Cooking

The cooking and processing of the grain for wheat flakes is similar in many respects to the processing of corn grits for cornflakes. Yet there are significant differences, one of the first of which is in loading the cooker with grain and flavor syrup. For wheat flakes, this is typically done by metering the grain from a weigh scale and pumping the flavor syrup through a delivery hose located at the cooker opening, in such a way that the two streams commingle as they fall into the cooker. Once the cooker is loaded and capped, it is common practice to rotate it through four to six revolutions to mix the syrup more thoroughly over the grain before turning on the steam.

It is not necessary to fill the cooker as full as a cooker used for corn grits. If it is too full at the end of cooking, the wheat mass may pack in the cooker and not dump unless the operator digs at it manually to loosen it.

A steam pressure of 15 psi is sufficient for cooking wheat. At this pressure, batches can be thoroughly cooked in 30–35 min rather than the 2 hr necessary for corn grits. At the end of the cooking period, the cook appears different from a corn cook. Corn rolls from the cooker as individual particles, which can be broken apart very easily. Wheat rolls from the cooker in rather tightly stuck-together balls ranging in size from that of golf balls to that of soccer balls. The pulling and tearing action required to pull these apart calls for a delumping system different from that for cornflakes.

Lump Breaking

Lump breakers for wheat flakes are commercially made units usually consisting of a rectangular steel frame in which one or two horizontally rotating shafts or drums are located. In a one-shafted machine, a matching and intermeshing comb is mounted on the interior wall of the frame. The rotation of the drum forces the material through the comb, crushing it into smaller pieces as it goes. In a two-shafted machine, the shafts usually rotate toward each other at differential speed, with projections on each intermeshing, thus tearing and crushing the material. Details on lump breakers are outlined in Chapter 3.

More than one lump breaker is usually necessary in a wheat flake line, with those closest to the cookers performing the coarse breaking and those farther downstream performing the finer crushing. Screening operations between lump breakers separate properly sized material from the stream and return oversized material for further size reduction.

Large volumes of air are drawn over the product and through the lump-breaking equipment. This is needed to cool the cooked wheat back down to near ambient temperature and to skin over the individual pieces, so that they have a nonsticky surface. Wheat overcooked or too high in moisture becomes very sticky and difficult to process down to the size of dryable and flakable grits.

The desired grit size for flaking is about 0.375 in. in diameter, with some grits as small as 0.125 in. and some up to 0.50 in. The grit size range determines the bulk density of the finished, toasted flakes and is the biggest factor determining the carton weight of the packed product.

Drying

Drying cooked wheat grits is not unlike drying corn grits. Wheat grits are somewhat more fragile, and care should be taken not to beat them around mechanically to the point of generating fines that must be removed before flaking in order to maintain a correct bulk density of the finished product. Moisture is more easily removed from wheat than from corn. Excessive dryer temperatures have a darkening effect, which carries right through to the finished product. This is true of all grains processed for breakfast cereals.

The moisture content of wheat grits from the dryer should be in the range of 16–18%. This is noticeably higher than that of cornflakes, one reason being that wheat flakes, unlike cornflakes, do not blister during toasting and therefore are not dependent on moisture content. Wheat grits are agglomerated particles made up of smaller particles stuck together. If they are too dry (12–13% moisture) they shatter into smaller pieces when flaked, resulting in a finished product with higher bulk density. If the grits are too high in moisture content (19–20%), they are too sticky and gummy for flaking. The flakes stick to the roll surfaces and are very difficult to scrape off.

Cooling and Tempering

After drying, it is important to cool the grits well below 110°F (43°C) before binning them to temper. If they are too warm going into a temper bin, the result is continued darkening, which carries over to the finished product.

The temper time for wheat grits is generally shorter than that for corn and does not exert as great an influence on the texture and appearance of the finished product, since wheat flakes do not blister during toasting.

Temper bins come in various sizes and shapes. Most consist of sides mounted over a slowly driven conveyor belt. They are loaded from an overhead conveying system in such a way that they can be unloaded on a first-in, first-out basis. This is most important. Usually some kind of raking device is needed to loosen the grits into individual pieces again.

After tempering, it is necessary once again to sift to remove fines and oversized grits. The latter are processed by a lump-breaking device and then resifted. The fines from this operation, plus those from other parts of the process and from packaging, are collected in a sanitary manner and added to subsequent cooks with new raw materials. If the amount of fines reworked in a cook is larger than about 100 lb in a 1,000- to 1,500-lb cook, a decrease in the consistency and the color quality of cooks and flakes is noted, to say nothing of the inefficiency involved.

Flaking

Good-quality wheat flakes can be made without any further pretreatment of the grits before rolling and toasting. Since the physical composition of the bran particles in the grits is so different from that of the endosperm, there are many points in the flake for uneven heating and drying during toasting. This characteristic improves the texture of the finished flakes, since it causes them to curl during toasting. A slight to moderate amount of curling is desirable, making the flakes appear more appetizing and interesting. If the flakes are perfectly flat, they lack interest, and they tend to lie flat in the carton, resulting in an excessively high net weight.

The curling points in wheat flakes are at interfaces of materials of different composition (bran and endosperm), slightly different moisture levels, or different rates of heat transfer.

The moisture content of finished toasted wheat flakes should be 1–3%. Essentially everything already said about the toasting of cornflakes also applies to wheat flakes. Nutritional fortification is discussed in Chapters 7 and 10.

RICE FLAKES

The processing of rice flakes differs in only minor ways from that of cornflakes and wheat flakes.

Formulation

Rice flakes can be made from head rice (whole grain) or second heads (broken pieces of whole kernels). From an economic standpoint, the latter is preferred. Whole grains are preferred for oven-puffed rice cereals, in which each kernel forms an individual piece of finished, toasted cereal, as described later in this chapter.

The broken grain size used is referred to in the U.S. standards for milled rice as second-head milled rice. The smaller sizes of broken kernels, referred to as brewers' rice, can also be used, but they may not be available, because of demand by the brewing industry.

A typical formula for rice flakes is as follows: second-head rice, 100 lb; fine-granulated sugar, 8–12 lb; malt syrup, 2 lb; salt, 2 lb; and water sufficient to yield a cooked product with a moisture content of about 28%. The flavor materials and water can be made up in a master batch sufficient for multiple cooks, and aliquots weighed for each cook.

Cooking

Cooking of rice grits is similar to that described for wheat flakes, except that the cooking time is 60 min at a steam pressure of 15–18 psi.

Moisture control and the sugar content are more critical in rice cooks than in corn or wheat cooks. Moisture contents much over 28% at the end of cooking render the cooked rice extremely sticky and difficult to handle through the cooling, lump-breaking, and initial drying phases. Likewise, sugar amounts much over 20 lb/cwt of rice cause excessive stickiness.

Lump Breaking and Drying

Lump breaking and drying in the processing of rice flakes are similar to these steps in the processing of wheat flakes. The properly sized grit for drying, tempering, and flaking is an agglomeration of many individual pieces of broken rice that have been stuck together by the cooking action. The identity of the individual pieces of broken rice is not lost, and it is maintained throughout cooking and subsequent steps.

Because of this grit construction, cooked rice grits must be handled more gently than corn grits. If handling is too rough, excessive amounts of broken grain pieces are abraded into useless fines, which have to be reworked or thrown out.

The drying time of rice grits is longer than that of wheat and more like that of corn. At the end of drying, a desirable moisture for flaking is 17%. If the moisture content is too low, the grit shatters during flaking, and small flakes are produced as a result. If the moisture content is too high, the flaking rolls gum up.

Cooling and Tempering

Cooling and tempering in the processing of rice flakes are similar to these steps in the processing of wheat flakes and cornflakes, with a minimum temper time of 8 hr being preferred.

Toasting

Toasting rice flakes requires more heat than toasting wheat flakes. The moisture content of the entering flakes and the heat of the oven must be such that the flakes blister or puff during toasting. If they do not, they become excessively hard and flinty. Blistering is usually accomplished by having the discharge end of the oven hotter than the feed end. As with wheat and corn, the moisture content of the finished rice flakes should be in the range of 1–3%.

EXTRUDED FLAKES

Extruded flakes differ from those made by the traditional process in that the grit for flaking is formed by extruding the mixed ingredients through a die hole and cutting off pellets of the dough in the desired size. A typical process using wheat as the basic grain is described here.

Formulation

A typical formula might be as follows: whole wheat kernels or whole wheat flour, 100 lb; sugar, 8–12 lb; salt, 2 lb; malt syrup, 2 lb; natural or artificial color as desired; and water sufficient for pellet formation and cooking.

Color is added to the formula for a definite reason. In the traditional process for cornflakes and wheat flakes as described in previous sections, mechanical working or shearing of the ingredients is kept to a minimum by the very nature of the processing steps. The resulting flakes have a brightness of color, called *bloom*, which makes them attractive and appetizing. When the ingredients are processed in extrusion systems, more mechanical working takes place, and the resulting flakes often appear dull in color, even slightly gray. This effect is increased if the formulation is low in sugars or does not contain malt syrup as a source of reducing sugars to participate in the Maillard browning reaction. A small amount of natural or artificial yellow color can overcome this effect.

Mixing and Extruding

If whole wheat kernels are used, soft red or white winter wheat is preferable. The extruder elements are set up in such a way as to knead or crush the wheat early in its path through the unit. Also early in the extrusion process, the flavor solution is added directly to the barrel of the unit by means of a metering pump. Sometimes it is preferable to hold out some of the water from the flavor solution and add it by means of a separate metering pump system. In this way the moisture content and consistency of the extruded pellet can be controlled without altering the flavor and color elements in the dough. Heat input to the barrel of the extruder near the feed point is kept low to allow the mixing of the ingredients before too much cooking and gelatinization start. Heat is applied to the barrel to accomplish the cooking. Very close control of temperatures is necessary. Chapter 6 and a section of Chapter 3 deal in greater detail with the extrusion process.

The last section of the barrel, directly behind the die, is usually cooler than the center sections. The dough, as it extrudes from the die, therefore remains in a compact form rather than expanded due to moisture flash-off. A knife rotating against the die severs the extrudate into pellets about 0.25 in. in diameter and 0.375 in. long. The moisture content of these pellets is usually 22–24%. If it is much higher than this, they are too gummy for good cutting. The moisture content can also be lower, depending on the equipment used. If it is possible to achieve complete cooking and flavor development at moisture contents in the range of 17–18%, so much the better, since this eliminates a drying step otherwise necessary before tempering and toasting.

Flaking and Toasting

Tempering the extruded dried pellets is only necessary if there is a great difference between their surface moisture and their internal moisture. In some cases it may be desirable for the outside to be drier or slightly skinned over, because the dry skin forms interfaces with the more moist portions, and the flakes twist and curl at the interfaces during toasting. If the moisture is very uniform throughout the pellet, flat, poker-chip-like flakes result.

Flaking and toasting the pellets is accomplished as explained in the traditional manufacture of cornflakes and wheat flakes. Nutritional fortification is dealt with in Chapter 7.

Gun-Puffed Whole Grains

Gun puffing of whole grains is a very interesting process. Two things are necessary for grain to puff—the grain must be cooked, and a large, sudden pressure drop must occur in the atmosphere surrounding the grain. As steam under pressure in the interior of the grain seeks to equilibrate with the surrounding, lower-pressure atmosphere, it is released.

Grains Used

Rice and wheat are puffed as whole kernel grains and marketed as puffed rice and puffed wheat cereals. Other cereal grain materials, such as corn and oats, also are used for puffing, but not as whole grains. These are discussed later as extruded gun-puffed cereals.

The rice used for puffing is either long-grain white rice or parboiled medium-grain rice. Other than normal milling to produce head rice, pretreatment of the grain before puffing is not needed. Milling should be such that the fat content of the rice is reduced to 0.5–1.5%. In batch guns, as manufactured by the Puritan Manufacturing Company, in Omaha, Nebraska, a small amount of water is added to the grain to generate steam for puffing.

The wheat used for puffing is generally hard wheat, preferably durum. In the trade such wheat is known as *puffing durum*, implying that the wheat has been specially sized and cleaned so that the range of individual kernel sizes is as small as good economics allows. Puffing durum carries a premium price because of this extra cleaning and sizing.

Pretreatment

Unlike rice, wheat requires a pretreatment step to avoid loosening the bran from the grain in a very ragged, haphazard manner, with some of it adhering and other parts being blown partially off the kernels.

One form of pretreatment is to apply about 4% saturated brine solution (26% salt). The water is used to generate steam for puffing. The salt toughens the bran during the preheating time and makes it adhere to itself better and become less fragile. The puffing action then blows the bran from the grain, resulting in a much cleaner-appearing and more appetizing puffed product.

Another form of pretreatment involves removing part of the bran from the grain altogether before puffing. This is accomplished by pearling, just as is done for the removal of rice bran in the manufacture of white rice. The grain is passed through a machine in which revolving, high-speed silicon carbide or Carborundum stones are mounted. These vary in grit size, depending on their position in the unit—the coarser being closer to the grain inlet, and the finer closer to the grain outlet. The bran coat is abraded off the kernels, and the bran particles are removed by air suction and deposited in a dust collector. About 4% of the kernel weight must be removed by pearling to make a good-quality puffed wheat.

Puffing

In batch puffing, single-shot guns (some of which are still in use) are heavy-walled steel vessels capable of withstanding pressures in excess of 200 psi, with an internal volume of 0.4–0.5 ft^3 (Figure 1).

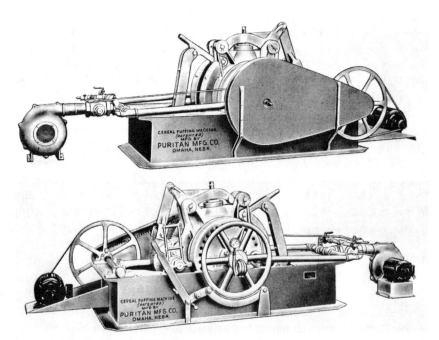

Figure 1. Single-shot cereal-puffing gun: drive side (top); operating side (bottom). (Courtesy Puritan Manufacturing Co., Omaha, NE)

The opening through which the gun is loaded with grain, and through which the grain is fired at puffing time, is closed and sealed with a lid operated by a system of levers and cams. At firing time this cam system allows the lid to be opened instantly. Heat is applied by means of gas burners with very hot flames impinging on both sides of the gun. During the heating cycle, as the gun rotates on a shaft mounted on each side of the gun body, the moisture in the grain and any added before gun loading is converted to steam. When the lid is opened to fire the gun, the internal pressure is released, and the puffed grain is caught in a continuously vented bin (Figure 2). A typical cycle for a batch firing is given in the accompanying box.

The operation of puffing guns can be dangerous if strict safety precautions are not adhered to. Operators should wear full safety face shields at all times. The firing noise is very loud, and so suitable ear protectors should be worn at all times. Diligence must be shown by operators to be sure that pressure gauges and safety valves are in proper working condition and that guns in operation are not allowed to exceed the desired firing pressure of 200 psi. Lid seals and locking cam devices must be kept in good working order at all times, or misfires are apt to occur. Misfiring of guns at unexpected times can cause very serious injuries.

Automation

Another type of single-shot gun is automatic in operation. In these guns, steam is injected into the gun body at 200 psi, and the time

---

Typical Cycle for Batch Puffing

Preheat the gun body to approximately 400–500°F (204–260°C).

Apply 0.8 lb water to 20 lb of rice or pearled wheat and distribute it uniformly over the grain.

Load the wetted grain into the gun.

Close and clamp the lid tightly with the cam and lever system.

Start the gun rotation.

Turn the gas flames on.

Monitor the gun until the internal pressure gauge reads 200 psi.

Stop the gun with the lid pointed in the appropriate direction for catching the puffed grain.

Turn off the gas.

Fire the gun by releasing the firing lever on the lid cam system.

Catch the puffed grain in a continuously vented bin (Figure 2).

then necessary to transfer heat to the grain and condition it for puffing is drastically reduced, from 9–12 min to as low as 90 sec. In most cases it is desirable to have about 100°F (55°C) of superheat on the 200-psi steam, to avoid producing excessively wet and soggy grain. Automatic puffing has been described by Maehl (1964).

Multiple-Shot Guns

In multiple-shot guns, multiple barrels are used on the same loading center and firing point. For any one barrel, the load, steam, and fire cycle is similar to that for the single-shot automatic just described, but several barrels are mounted on a slowly rotating wheel so that each passes the load and fire positions at the correct time, with steaming taking place in between. Such automatic batch guns lend themselves to modern-day electronic process controls, which help remove the danger to operators through human carelessness.

Several methods of puffing grain continuously have been developed and patented (Haughey and Erickson, 1952; Perttula, 1966; Tsuchiya et al, 1966; Paugh, 1975; Dahl, 1976). These generally involve admitting grain to an already steam-pressurized puffing chamber by means of a special valve and subsequently releasing the thorougly heated grain

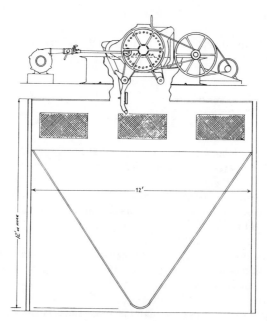

Figure 2. Vented bin for puffed grain, with a single-shot gun in shooting position. (Courtesy Puritan Manufacturing Co., Omaha, NE)

to the atmosphere through an orifice without loss of pressure in the chamber. One such system is diagrammed in Figure 3.

Final Processing

After the puffed grain is caught from the puffing guns, it must be processed further before packaging. The first step is to screen the product to remove unpuffed kernels, bran and dust particles, and small, broken puffed kernels. The moisture content of the grain from most guns is 5-7%, which is too high for a crisp cereal. Accordingly, the

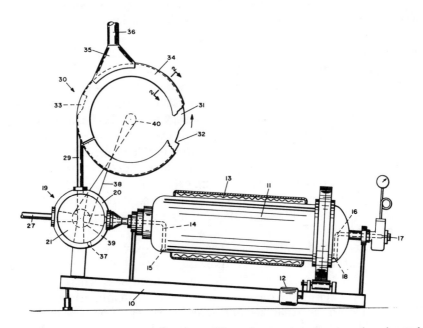

Figure 3. Apparatus for continuous puffing. Successive charges of preheated grain and pressurized steam are directed from a gravity-fed hood (35) by a centrifugal thrower (30) (which has two receiving pockets [32 and 33]) through a pipe (29) into one or another of the passageways of a rotary charging valve (21), from which they are directed successively through a fixed pipe (14) into a rotating pressurized chamber (11) equipped with electrical heating coils (13). The grain is further processed as it is moved to the lower end of the chamber by tumbling and mixing in the pressurized steam environment of the chamber. At the lower end, it is puffed as it leaves the high-pressure chamber through an orifice (17) into atmospheric pressure. A typical puffing chamber is 30 in. in diameter, 50 in. long, and has an orifice 0.5 in. in diameter. The centrifugal thrower is 36 in. in diameter, turning at 72 rpm, and the charging valve rotor is 16 in. in diameter, turning at 36 rpm. (Reprinted from Dahl, 1976)

second step is to dry the grain down to 1–3% moisture. Since the grain is very porous, it takes up moisture very rapidly and easily, so that package materials with good moisture barrier qualities are needed.

Extruded Gun-Puffed Cereals

Extruded gun-puffed cereals originate from flours and not from whole grains. The cooking usually takes place in extruders. The cooked dough is then formed into the finished shape by means of extrusion through a die, with cooking and extrusion sometimes a one-step instead of a two-step process. An example of this type of product is General Mills's Kix, which was one of the first such products on the market.

Mixing and Extruding

A typical sequence is as follows. First, the basic dry materials—flours, starches, and heat-stable microingredients—are premixed together in a uniform blend, and a solution of sugar, salt, malt, other flavors, color, and water is made up. The dry material is fed to a cooking extruder; the flavor solution is added through the barrel of the first section of the extruder, and more water is added separately. If the amount of flavor solution is kept in constant balance with the amount of dry blend, a constantly flavored and colored dough is produced. The separate addition of water can be adjusted as needed for extruder control and to compensate for minor fluctuations in the moisture content of the raw materials.

After the cooked dough exits the cooking extruder, it is fed to a forming extruder, which is usually controlled at noncooking temperatures below 160°F (71°C). The extruded shape here may not be the exact shape of the finished gun-puffed piece in miniature, because of differences in the expansion of its parts during puffing. The moisture content of the extruded cooked shapes is usually in the range of 20–24%.

Drying and Tempering

The next step after extrusion is drying and tempering, usually to a moisture content of 9–12%. The shapes are then gun-puffed as already described for whole grains. No pretreatment is needed, although with nonautomatic single-shot guns it may be necessary to add 2–4% water to assist in generating steam pressure. Most of these shapes puff adequately under pressures in the range of 150–200 psi.

Final Processing

After puffing, the finished product must be screened to remove unpuffed shapes, broken pieces, and dust, and then it is dried to a moisture content of 1–3%, the normal level for a finished cereal.

Many gun-puffed extruded products are sugarcoated as well as nutritionally fortified before packaging. Sugarcoating and nutritional fortification are covered in detail in Chapters 7 and 10. Additives such as marshmallow bits in interesting shapes and colors may also be included.

Shredded Whole Grains

The grain used in whole kernel form for shredding is primarily wheat. White wheat produces shredded wheat biscuits that are light in color with a golden brown top crust and bottom when properly baked. Red wheats can also be used for shredding, but the shreds are more gray, and bran specks stand out more, because the bran is darker to start with. Rice, corn, and other grains also can be used for shredding; they are covered in the next section of this chapter.

Cooking

The process for shredding wheat begins with cleaning the wheat of all sticks, stones, chaff, dust, other grains, and foreign material. Once cleaned, the wheat is cooked in batches in excess water at slightly below the boiling point under atmospheric pressure. Cooking is achieved and stopped when the very center of the kernel endosperm turns from starchy white to translucent gray, which usually requires 30–35 min.

The cooking vessels usually have horizontal baskets that rotate within a stationary housing and are of sufficient size to hold 50 bu of raw wheat (about 3.5 ft in diameter and 8 ft long). They are equipped with a water inlet and drains. The heating medium for the water is steam injected directly into the water inside the cooker. Water-temperature-sensing probes control the flow of steam to maintain the desired cooking temperature. The moisture content of the cooked grain at the end of the cooking cycle is 45–50%.

Cooling and Tempering

After the completion of the cooking cycle, the water is drained from the cooker, and the wheat dumped and conveyed to cooling units. These can be vertical, louvered units through which air is drawn, or they can be horizontal, vibratory, perforated pans through which refrigerated air is circulated. Whichever is used, the objective of this step is to surface-dry the grain and cool it to ambient temperature, to stop the cooking process.

After cooling, the wheat is placed in large holding bins and allowed to temper for up to 24 hr before shredding. This holding time allows the moisture in the kernels to fully equilibrate. The kernels become more firm, probably because of retrogradation of the starch (Jankowski and Rha, 1986). This firming of the kernels is vital for obtaining shreds of good strength for cutting and for handling of the unbaked biscuits. If the holding time is insufficient, the shreds will be crooked rather than straight, as well as gummy and sticky, and cannot be cut properly.

Shredding

In the shredding operation, the wheat kernels are squeezed between two rolls—one with a smooth surface, the other grooved. Although simple in concept, in actual practice this operation requires a good deal of skill and attention. The wheat is squeezed into the grooves of the grooved roll. Positioned against this roll is a comb, each tooth of which fits into one of the grooves in the roll. As the roll revolves with its grooves filled with cooked wheat, the comb teeth pick the wheat shred out of the groove. The shreds are laid down on a conveyor under the rolls running parallel to the shredding grooves. In the traditional process, each pair of rolls forms one layer of the finished biscuit.

Many different variations in rolls and grooving are used in the cereal industry today. Typical single-layer rolls are 4 in. long and 5 in. in diameter, with up to 20 grooves per linear inch of roll length. Some also have grooves running across the roll. The resulting cross shreds provide extra strength in the web.

Other variations include the shape and dimensions of the grooves themselves. Some are U-shaped, and others are V-shaped. They are generally 0.017–0.022 in. in width and depth, these being the determining factors of the texture and appearance of the finished biscuit. Grooves of different dimensions can be used on the same shredding line to maintain proper weight control of the unbaked and baked finished cereal. The roll size also varies depending on the output desired, the type of product being run, the power available to drive the rolls, and the composition and wear qualities of the roll steel.

With the two rolls touching each other, frequently revolving at different speeds, there is bound to be wear of the roll surface. As this occurs, the groove dimensions change, and so do the shred size, weight, and texture. The speed differential is usually in favor of the grooved roll by 4–20%, since the material being shredded has a tendency to stick better to the faster roll.

Roll surface temperatures at optimum shredding are in the range of 95–115°F (35–46°C). Sometimes it is necessary to water-cool the rolls to optimize the roll temperature.

Forming Biscuits

If each pair of rolls lays down only one layer of shreds, 10–20 pairs of rolls are needed to produce a sufficiently deep web of shreds to form large shredded wheat biscuits. However, it is also possible and may be economic to use fewer and larger-capacity rolls operated so as to lay down the equivalent of several layers each. Bite-sized products require fewer layers. After the web of many layers of shreds reaches the end of the shredder, it is fed through a cutting device to form the individual biscuits. The cutting edges of the cutter are usually dull rather than sharp, so that the cutting action is in part a squeezing, which compresses the shreds and makes them stick to each other. This forms a crimped joint, which holds the shreds together in the biscuit form.

Baking

The individual biscuits are then baked in a band or continuous conveyor-belt oven. This is zoned and controlled so that the major heat input to the biscuits is in the first few zones, a rise in biscuit height occurs in the middle zones along with moisture removal, and color development and final moisture removal occur in the last few zones. The temperatures are in the range of 400–600°F (204–315°C). The moisture content of biscuits going into the oven is usually about 45%, and the final moisture content out of the oven is about 4%.

Shredded wheat is susceptible to oxidative rancidity, so that it must be protected by antioxidants for a shelf life of more than a few weeks. Antioxidant technology is dealt with in Chapter 10.

Extruded and Other Shredded Cereals

Raw materials used for extruded and other shredded cereals are wheat, corn, rice, and oats—alone or in mixtures. These basic grains can be in the form of whole kernels, parts of kernels, or flours, depending on the product being made and the form of cooking used before shredding. Other ingredients also can be incorporated (e.g., starches, sugar, corn sweeteners, malt, salt, color, flavors, and vitamin and mineral fortification mixes), just as in extruded flaked or puffed products.

Pressure Cooking and Extrusion

The precooking that takes place before shredding may be either pressure cooking, as described for flakes, or extrusion cooking. If it is done in normal pressure cookers, there is a major difference in handling after cooking for shredding as opposed to flaking. Unlike flaking, in which flakes of uniform size are desired, shredding does not require any attempt to keep the size of the individual pieces within exact, narrow limits. It is only necessary to reduce the cooked lumps sufficiently in size that they can be cooled and can be fed uniformly to the nip of the shredding rolls. Extrusion cooking allows better control over the size of individual pieces, as determined by the cut at the die face of the extruder.

The moisture content of the cooked dough pieces for shredding is much lower than that of cooked whole wheat (25–32% for either pressure-cooked or extruded wheat vs. 45–50% for wheat atmospherically cooked for shredding). Normally no drying step is needed between cooking and shredding, but cooling is required, to remove the heat of cooking and stop further cooking from taking place. Furthermore, in the case of pressure-cooked formulas, tempering is required after cooling. This allows the material to equilibrate in moisture content. The tempering time can range from 4 to 24 hr. Tempering is usually not needed for extrusion-cooked formulas.

Shredding

Shredding takes place much as already described for whole grain shredding. However, where wheat, corn, or rice with sugar, salt, and malt make up the formula, heat is generated in the shredding rolls at a much faster rate. Water-cooled rolls are essential unless roll pairs can be switched very frequently, such as every 1–4 hr. The heat is generated because the formula has a lower moisture content (30% vs. 45% for whole wheat shredding) as well as a tougher cooked mass at the time of shredding, compared with whole-cooked wheat.

Shreds from cooked dough are almost exclusively formed into bite-sized finished products, requiring many fewer layers of shreds. In corn and rice products, the number of layers can be as few as two to four, and in wheat and oat products four to eight. Usually these products are made with cross grooves on the shredding rolls. These aid in adding extra strength to the web of shreds and produce finished products with a more uniform appearance and fewer "crippled" or distorted pieces.

Cutting

Cutting is performed very much like the cutting of whole grain products. The cutting edges are dull and squeeze the web of shreds together. After cutting, these products are usually baked while still tied together in sheet form. They tend to shrink in length and width during baking, and the shrinking helps to weaken the cut point and makes it very easy to break the individual pieces apart after baking.

Baking

Shredded wheat and oat products are usually baked in long, continuous band ovens, with bake times of 1–4 min. If the products are cut all the way through the web of shreds, they can be baked in a fluid-bed toaster in about the same time. The moisture content of the finished product should be in the range of 1.5–3.0%. Antioxidant treatment is usually required, to protect the product against rancidity and ensure a reasonable shelf life.

Corn- and rice-based shredded cereals are toasted or baked in a more specialized manner. These shreds must be puffed or opened up during toasting. They become extremely hard and flinty in texture if they are not puffed. To achieve the puffing, the first part of the oven is usually at a lower temperature and dries the shreds. In the last half of the oven is a section of extremely high temperature (550–650°F, or 288–343°C), causing the rapid flash-off of the remaining moisture, which puffs the shreds.

Oven-Puffed Cereals

Oven-puffed cereals are almost exclusively made from rice or corn or mixtures of these two grains. These two grains inherently puff in the presence of high heat when the moisture content is correct, whereas wheat and oats do not.

Formulation and Cooking

Usually medium-grain rice is the starting material for oven-puffed rice. It is pressure-cooked with sugar, salt, and malt flavoring for about an hour at 15–18 psi. A typical formula is as follows: medium-grain white rice, 100 lb; sugar, 6–10 lb; salt, 2 lb; malt extract, 2 lb; and water sufficient to yield cooked rice at 28% moisture.

After cooking, the rice is conveyed through a cooling and sizing operation, which removes the heat of cooking and returns the rice to ambient temperature. It also breaks agglomerates into individual kernels.

Drying and Bumping

Drying is a two-stage process, with additional steps between the first and second drying. First, the rice is dried to reduce the moisture content from 28% to about 17%. Then it is tempered 4–8 hr, or long enough for good moisture equilibration. After tempering, it is bumped, that is, run through flaking rolls to slightly flatten the kernels but not make thin flakes out of them. Bumping presumably creates fissures in the kernel structure, which promote expansion at high oven temperatures. The thinner dimension also allows for faster heat penetration. Bumping is essential for proper puffing in the heat of the oven.

After bumping, the rice is dried a second time, to reduce the moisture content from 17% to 9–11%. This second drying is needed for good oven puffing, an operation that requires a proper balance between the grain moisture content and the oven temperature.

Oven Puffing

Generally, oven puffing is characterized by extremely high oven temperatures (550–650°F, or 288–343°C) in the latter half of the oven cycle. Final toasting and puffing are accomplished in about 90 sec in rotary flake-toasting ovens or other fluid-bed ovens. After oven puffing, the cereal is cooled, fortified with vitamins if used, and frequently treated with antioxidants to preserve freshness.

Granola Cereals

The major raw material used to make a granola cereal is rolled oats, either regular whole-rolled (the old-fashioned type) or quick-cooking oats. Mixed with the oats are other interesting raw materials, such as nut pieces, coconut, brown sugar, honey, malt extract, dried milk, dried fruits (raisins, dates, etc.), water, and vegetable oil (Bonner et al, 1973). Spices such as cinnamon and nutmeg can also be added.

The water, oil, and other liquid flavorings are made into a suspension. The oats are blended with the other dry materials. The liquids and dry blend are mixed together in the proper amounts, and the wetted mass then spread in a uniform layer on the band of a continuous dryer or oven. Small volumes can also be produced by spreading the wetted mass in a uniform layer on baking pans for batch baking.

Baking takes place at temperatures in the range of 300–425°F (149–218°C) until the mat is uniformly toasted to a light brown and moisture reduced to about 3%. After toasting, the mat is broken up into chunky pieces.

Since most of the granolas on the American market are sold as "natural," they are not treated with antioxidants. Nor do they contain artificial flavors or colors.

Extruded Expanded Cereals

The basic cereal processes of flaking, gun puffing, and oven puffing are methods of converting raw, dense grain (48 lb/ft^3) into friable, crisp, or chewable products suitable for human food, with a bulk density in the range of 4–10 lb/ft^3. Extrusion is merely another technology operating on a continuous basis for the conversion of basic dense grain formulations into light and crisp products that humans find enjoyable and nutritious. Typically, however, the grain in the formula for extruded expanded cereals is flour or meal rather than whole or broken kernels. Cooking with water and flavor materials is accomplished by means of a cooking extruder or the cooking section of a cooking-expanding extruder.

Once the formulation has received its cooking, either in the cooking section or in its own cooking extrusion unit, it is expanded when the moisture in the formula (whether natural or added) is released from a zone of elevated temperature and pressure to ambient conditions. Holes in the die at the end of the extruder control the shape of the finished cereal pieces once they are cut. The cutting of the expanded or expanding extrudate is usually done by a knife rotating on the outer face of the die. The extruded expanded pieces can be sugarcoated or colored and flavored to produce a variety of products for various tastes.

One early process for the production of an extruded expanded cereal was described by Fast et al (1971) and another by Rosenquest et al (1975). Many more exist. The use of extrusion for continuous cooking is covered in Chapter 3, and Chapter 6 is devoted in detail to extrusion in breakfast cereal manufacturing.

References

Bonner, W. A., Gould, M. R., and Milling, T. E. 1973. Ready-to-eat cereal. U.S. patent 3,876,811.

Dahl, M. J. 1976. Apparatus for continuous puffing. U.S. patent 3,971,303.

Fast, R. B., Hreschak, B., and Spotts, C. E. 1971. Preparation of ready-to-eat cereal characterized by honey graham flavor. U.S. patent 3,554,763.

Haughey, C. F., and Erickson, R. T. 1952. Puffing cereal grains. U.S. patent 3,876,811.

Jankowski, T., and Rha, C. K. 1986. Retrogradation of starch in cooked wheat. Starch/Staerke 38(1):6.

Maehl, E. F. 1964. Cereal puffing apparatus. U.S. patent 3,128,690.

Matz, S. A. 1959. Chemistry and Technology of Cereals as Food and Feed. AVI Publishing Co., Westport, CT.

Paugh, G. W. 1975. Continuous puffing method. U.S. patent 3,908,034.

Perttula, H. V. 1966. Feed valve. U.S. patent 3,288,053.

Rosenquest, A. H., Knipper, A. J., and Wood, R. W. 1975. Method of producing expanded cereal products of improved texture. U.S. patent 3,927,222.

Tsuchiya, T., Long, G., and Hreha, K. 1966. Method and apparatus for continuous puffing. U.S. patent 3,231,387.

Chapter 3

Unit Operations and Equipment I: Blending and Cooking[1]

ELWOOD F. CALDWELL
ROBERT B. FAST
CHARLES LAUHOFF
ROBERT C. MILLER

In this and the next three chapters, we review the major unit operations of most processes by which ready-to-eat breakfast cereals are produced. We also discuss examples of the types of equipment used in these processing steps. The operations discussed in this chapter are blending, cooking, lump breaking, and sizing.

Blending

After the ingredients for the cereal to be produced have been selected, it is necessary to make a uniform blend of them before further processing begins. Blending can be broken down into three phases: dry blending of the dry ingredients, liquid blending of the liquid ingredients, and combining liquid and dry ingredients before further processing.

DRY BLENDING

The blending of grains or grain pieces and other dry materials, such as flours or starches, is most commonly accomplished in ribbon blenders,

[1]Principal contributor of the sections on blending and flavoring and on sizing and lump breaking, Robert B. Fast; on batch cooking, Charles Lauhoff; and on continuous cooking, Robert C. Miller.

sometimes called horizontal batch mixers. One such blender is shown in Figure 1. Two ribbons of the agitator face in opposite directions. This reverse-spiral design performs uniform, high-capacity blending, with the mix held in constant suspension (Anonymous, 1981). Blenders of this type are made of either mild or stainless steel, stainless being preferred where clean-in-place cleaning is used. Gasketed, hinged covers fit tightly to minimize dust leakage. These blenders are sized for each particular cereal batch. Usually one batch per blender is sized for one batch cooker load.

No bolts, screws, pockets, or ledges should be present on the inside of the blender. These are only places where filth can lodge, so they should be avoided. All bearings should be outboard of the blender shell, so that no grease can migrate to the inside of the blender and contaminate the product. In operation the batch size should be such that the level of ingredients during blending does not exceed the height of the ribbon itself, to prevent nonuniform blending. Overloading for higher output is a common cause of incomplete blending.

Figure 1. Ribbon blender (Eureka Sanimix). (Courtesy S. Howes Co., Inc., Silver Creek, NY)

Loading and Discharging

The mixers are loaded in any number of ways, the simplest and cheapest being to open the lid and dump the ingredients in by hand. This is often the worst method, creating large amounts of dust in the surrounding area. An improvement over hand dumping is the use of a bag dump station fitted with a hood and suction to remove dust to a dust collector. A still better method is loading from a receiver (above the mixer) in which all the ingredients have previously been placed. Loading the blender from a receiver is the fastest way to load and therefore plays a significant role in how many batches per hour can be turned out by the blender. The cycle time per blender is the total time of loading, blending, and discharge.

A still more sophisticated way to load a blender is to mount the whole unit on load cells and use it as both a batch weigh vessel and a dry mixer. Dry ingredients can then be fed through multiple openings in the blender lid. Such a setup lends itself to feeding control by a programmable logic controller.

Discharge of the blender is accomplished through a discharge hole and gate assembly on the bottom of the mixer. The ingredients can be discharged to any suitable conveying system to the next step in the process. They can also be discharged to an integral surge bin, in a continuous processing setup.

CONTINUOUS BLENDING AND MIXING

In continuous processes, blending the dry material and then mixing with liquids can most often be accomplished in a continuous blender. These units most commonly consist of one or two shafts mounted in a horizontal housing. The dry materials to be blended are fed in at one end; liquids, if added, are injected through ports in the housing downstream of the dry feed; and the finished blend is discharged at the opposite end from the feed. Figure 2 shows such a unit, which is single-shafted. They are available in carbon or stainless steel and are designed to blend syrups, oils, and flavor solutions into dry materials.

Paddle design can vary. Solid paddles are available. Most commonly all paddles are mounted on the shafts in such a way that their pitch forward or backward may be varied to ensure proper retention time and uniformity of mix. Loop paddles are also employed if a more gentle mixing action is needed. These mixers are fitted with clean-out doors, which swing down to provide easy access to the interior surfaces for cleanup.

LIQUID BLENDING

By far the most prevalent flavor materials used in breakfast cereal manufacture are barley malt extract and dry or liquid sugar. In addition, salt, other flavor materials, and heat-stable vitamins that are water-soluble may be included in a liquid flavor solution. Portions of the solution are then blended with the dry grain mix before further processing.

Barley malt extract has a flavor that blends well with that of toasted grain, and it is bland and inoffensive for a food eaten as the first food of the day. It has been used in breakfast cereals almost since their inception. However, it is a very difficult material to weigh and move, being composed of 80% solids, very viscous, sticky, and unpumpable. It is distributed in 55-gal drums, and discharging from the drum is best accomplished by inserting a spigot in the bung on the drum end and allowing the extract to flow by gravity.

Malt extract is most often used in amounts of 1-4 lb per hundredweight of grain. In order to disperse it uniformly over the grain it must be diluted with water. This is most often done in batch kettles or tanks fitted with agitators, of the type shown in Figure 3. Weighed portions, most often for several batches, are placed in the kettle with weighed or measured amounts of water. Equipment to measure batches

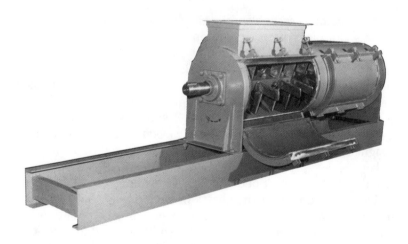

Figure 2. Continuous blender, single-shafted (Dynamic Mixer). (Courtesy S. Howes Co., Inc., Silver Creek, NY)

of liquid ingredients volumetrically must be carefully engineered to avoid inconsistent deliveries of amounts from batch to batch. Weighing liquid ingredients is preferable. In continuous systems, either a modern-day volumetric or a loss-in-weight gravimetric measuring system may be used.

Another type of agitator for mixing flavor solutions is shown in Figure 4. This type of mixer is most often mounted on the edge of an open tank. The action of the propeller blade provides the necessary mixing and agitation to blend the malt, water, sugar, salt, and other ingredients. More sophisticated versions can be had, in which the propeller-type blades are mounted through the side wall of specially made tanks.

In all of these liquid flavor applications care needs to be taken in the selection of the materials of construction of the vessels and piping.

Figure 3. Batch kettle with double-arm agitator. (Courtesy Lee Industries, Inc., Philipsburg, PA)

Surprisingly, some of these mixtures can be corrosive. Stainless steel should always be used, and grade 316 stands up much longer in service than the normal 304.

BLENDING DRY AND LIQUID MIXES

The third blending step is the blending of the liquid flavor solution uniformly over the blend of grain and other dry materials. This can be accomplished in single- or double-shafted continuous mixers or in simple cut-flight screw conveyors. There is typically agitation in the cooking step that follows, so blending of liquids and drys is usually accomplished simply and easily.

Cooking

Cooking is a necessary step in every process by which ready-to-eat breakfast cereals are produced. In addition to the development of

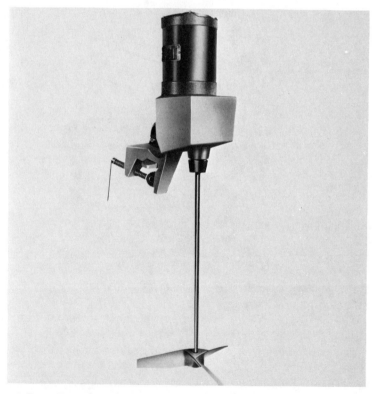

Figure 4. Propeller-type mixer for mounting on the edge of an open tank (Lightnin Mixer). (Courtesy Mixing Equipment Co., Rochester, NY)

desirable flavors (Harper, 1981) and nutritional benefits (Björck and Asp, 1984), the cooking step creates the physical properties necessary for the development of product texture—primarily by gelatinization of the starchy grain fractions. This textural development not only is important for aesthetic acceptability but is required in order to make the product edible. The structure is converted into an edible state by some form of expansion or development of cellularity.

Structural development is generally caused by the evolution of water vapor from rapid heating, in the case of toasting or oven puffing, or by the sudden release of pressure from the product in a superheated state, as in gun puffing or extrusion puffing. To respond to the vapor pressure, the product must be tough, elastic, and more or less homogeneous—a colloidal gel created by starch cooking (Williams et al, 1977).

Starting with a basic cooking step, the flow diverges through several process paths, depicted in Figure 5, to emerge from final texturization as the distinctly different cereal products already discussed in Chapter 2—flaked, shredded, oven- and gun-puffed, and direct-expanded or extrusion-puffed. In this section, we deal with the particular style of cooking practiced for each of these products.

BATCH VS. CONTINUOUS COOKING

Cereals are traditionally cooked by batch methods—by boiling whole grains in water for shredded wheat, for example, or steam pressure cooking (Tressler and Sultan, 1975). For many products, these methods are still in common use—it is sometimes difficult to duplicate a traditional product with a new process. However, some products have been duplicated with newer continuous processes, and many new products begin with the proposition that they are to utilize more modern methods.

Continuous processing offers several economic advantages, including potentially reduced energy cost, as a result of lower-moisture cooking and more efficient energy transfer; more efficient use of plant space, since the equipment used in continuous processes is usually smaller; reduced labor cost, because of compatibility with automation; and, sometimes, reduced ingredient cost, where mechanical manipulation can compensate for lower-quality materials, allowing the use of cheaper ingredients.

Perhaps of even more importance is that continuous processes permit better control. They typically operate with a shorter residence time than batch processes, constantly providing opportunities for corrective

action when a flaw is detected. Since the process stages preceding or
following the cooking step (such as flaking, toasting, or shredding) are
often continuous, the mismatch in product flow must be absorbed by

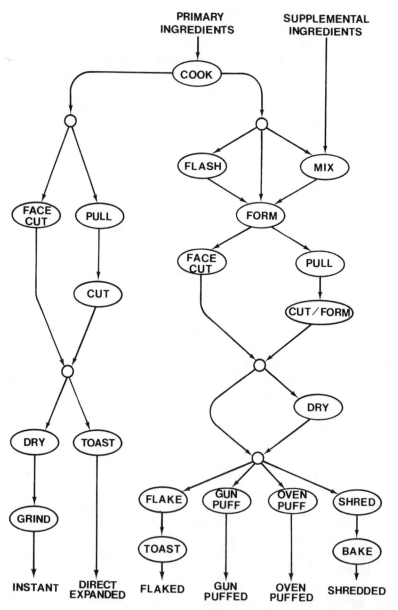

Figure 5. General process flow diagram for breakfast cereal manufacturing.

a "surge" area—the cooked product must be held for a period of time before it enters the next stage of the process. This not only requires extra equipment and plant space but causes uneven processing. Holdup time may be minimized to an extent by a "continuous-batch" flow system, in which the flow from a series of cookers is diverted to a series of next-step processes by carefully controlled batch scheduling, but it cannot be eliminated. In the worst case, the holdup time is equal to the cooking time plus any transfer time between processes.

The advantages of product uniformity in continuous processes are offset to a degree, however, by variations in residence time within the cooking process, created by the conveying mechanism of the process. Ideally, in batch processes, every element of the product is cooked for exactly the same time and, with proper agitation, receives about the same degree of cook. In continuous cookers, this uniformity could be matched only if the product passed through the cooking vessel with no mixing of the older and the newer feed material, that is, if plug flow were achieved—an impractical ideal. (The plug flow model is based on the movement of a rigid plug of material through the cooker.) In all standard processes, a degree of back-mixing occurs, caused by the forced movement through the cooking vessel. In the most extreme case, complete mixing creates a very wide residence time distribution.

Residence time distribution may be measured by adding a pulse of tracer material to the feed stream and then analyzing the product to determine the tracer concentration at time intervals. In plug flow, the tracer emerges in its original concentration, as a pulse. In all other cases, the tracer concentration reaches a maximum value and then tapers off as it is diluted by fresh material. Complete back-mixing, in contrast to plug flow, causes an initially high concentration that slowly dissipates in an exponential fashion. This has been well described by theory (Levenspiel, 1962) as

$$C = C_0 e^{-t/t_{av}} \tag{1}$$

with

$$t_{av} = V/v \quad \text{and} \quad C_0 = v_t/V,$$

where C = tracer concentration at time t after injection, C_0 = initial tracer concentration (at $t = 0$), t_{av} = mean residence time of the process, V = filled volume of the process, v = volumetric flow rate through the process, and v_{tr} = volume of tracer added at $t = 0$. Integration of the exit concentration curve generates the residence time distribution, which in a process with back-mixing becomes

$$f = 1 - e^{-t/t_a} ,$$ (2)

where f = fraction of flow through the system at time t. The residence time distribution for plug flow, by contrast, is a simple step function, with all of the tracer emerging at the same time. In real cooking processes, the actual residence time distributions lie between these extremes, as illustrated in Figure 6.

Extruders as Continuous Cookers

Single-screw devices, commonly used as continuous cookers, convey the product by shearing it and inherently cause some back-mixing. In simple cases, the residence time distribution of single-screw extruders has been described by theory (Tadmor and Klein, 1970), which provides a measure of comparison to real situations in experimental determinations (Bruin et al, 1978; Davidson et al, 1983).

Modified screw-type devices, which include ribbon agitators, cut-flight screws, and paddle mixers, produce a greater degree of back-mixing, shifting the residence time distribution toward that of complete back-mixing. Intermeshing twin-screw extruders shift the residence time distribution toward that of the plug flow ideal, particularly those with counterrotating screws, which create a positive-displacement

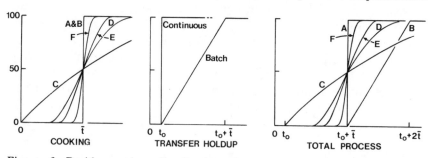

Figure 6. Residence time distributions in continuous and batch cooking processes (schematic)—percentage of the product through the process at time t in the cooking process alone (left), in the transfer time between cooking and a subsequent process (center), and in the combined cooking and transfer times (right). $\bar{t}$ is the median residence time, or the time required for 50% of the product to emerge; t_0 is the minimum conveying time. The curves describe different processes: A = continuous plug flow; B = batch cooking; C = 100% back-mixing; D = single-screw extruder; E = corotating twin-screw extruder; and F = counterrotating twin-screw extruder.

pumping mechanism (Janssen, 1978). Corotating twin-screw extruders produce a residence time distribution intermediate between that of counterrotating mechanisms and that of single-screw mechanisms.

The practical implication of residence time distribution involves product quality—if the distribution is broad, a portion of the product may receive excessive or insufficient cooking, even though the average heat treatment is correct for the product. This manifests itself as a particular problem when the formula contains heat-sensitive materials, such as milk, which can develop a scorched flavor and color. There is a trade-off between the virtues of mixing and of residence time distribution, which must be addressed in the specification of a cooking process. These factors are often mistakenly disregarded in scale-up procedures, but they should be taken into account by measurement of tracer exit concentrations.

BATCH COOKING

The standard batch cooker is of the rotary type and essentially consists of a horizontal pressure vessel that turns on its axis (Figure 7). The capacities of these cookers are typically in the range of 1,200–2,000 lb per batch, although both smaller and larger units are occasionally seen, the former primarily in the laboratory. Several manufacturers offer proprietary designs in various sizes. Fabricating firms can also build such vessels to a user's design. The disadvantage here is that the fabricator, as opposed to the proprietary manufacturer, builds to the American Society of Mechanical Engineers (ASME) Code and customer specifications. The efficiency and efficacy of the user's design are then the total responsibility of the buyer, including the elements of machine design that enter into the completed cooker.

Batch cookers find their greatest use in the cooking of whole grain cereals such as cornflakes, wheat flakes, and rice flakes, made directly from the preprocessed grain. They have also been successfully used for the cooking of wheat-based cereals such as bran flakes or shredded wheat, for which the wheat berry has been coarsely ground.

Where batch cookers continue to be used despite the availability of continuous process systems, one or more of the following circumstances usually prevails: 1) no alternative method, such as continuous or extrusion cooking, has been found to duplicate the taste and appearance of a traditional cereal product, such as cornflakes; 2) continuous cooking has been abandoned because of sanitation and maintenance problems; or 3) process control technology has been so developed that the only difference between batch cooking and other cooking processes is the load-unload cycle time.

Loading of Batch Cookers

Batch cookers are normally top-loaded through the charge-discharge opening. Standard loading mechanisms for raw materials are either conveyors or tram-mounted tanks. Conveyors consist of a trough with gates and retractable legs or boots over each cooker. Tram-mounted tanks run on a monorail over the cooker line and typically service each of several cookers in succession. These too have a gate and boot for filling the cooker.

Steam Injection and Venting

Steam-injection systems vary according to the vessel manufacturer and the user's preference. Steam can be injected through nozzles in the head ends or through a central internal manifold. Efficient batch cooking in pressure vessels requires that the cooker be bled or vented during the cooking process. This is essential so that additional new

Figure 7. Rotary batch pressure cooker with automatic controls. (Courtesy Lauhoff Corp., Detroit, MI)

steam can be constantly circulated through the material being cooked. It is also essential at the end of the cooking process to remove all steam from the vessel before opening it to discharge the cooked material. Venting can be accomplished by manual or automated valve systems and can include a vacuum system for the removal of steam.

Flavor Injection

Flavoring can be introduced into the product being cooked in three ways. Starting with the most common, these are 1) direct injection through the steam line, either into the center tube or through ports in the cooker heads; 2) bulk introduction through the charge-discharge opening; and 3) premixing with the product to be cooked (for pelletized products, flavoring may be introduced in the forming or preforming operation).

Lid Locking

Most cookers have traditionally been equipped with some type of quick-actuating closure. This is to afford quick emptying of the vessel contents, since the material can continue to "cook" if it remains in a hot, closed vessel, even though the steam is shut off.

There are numerous designs for quick-actuating closures, some proprietary in nature. The ASME Code (ASME, 1989) covers the design conditions that govern these closures. The user should be sure that any manufacturer's design meets the ASME criteria. However, the ASME recognizes that it is impractical to write detailed requirements to cover the multiplicity of devices used for quick access or to prevent negligent operation or the circumventing of safety devices.

Mixing

Mixing or agitation is required during the cook to ensure thorough cooking and uniform flavor. Vessels that are either conical or spherical have a geometry that promotes mixing, since the material gravitates to the center. Vessels with straight shells require internal flights or wings attached to the shell to first agitate the product and later move it to the discharge opening.

Cleaning and Maintenance

In a modern batch cooker the interior surfaces are highly polished, and to a certain extent the interior of the vessel is self-cleaning. However, the vessel should still be thoroughly washed and cleaned at reasonable intervals. Whether cleaning is performed weekly or daily depends on the type of material being cooked. Some materials tend to adhere to surfaces much more than others.

The batch cooker, although simple in concept, is nevertheless both a pressure vessel and a machine. Consequently it requires special attention in certain respects that do not typically apply to other types of process equipment: 1) regular inspection and replacement of seals, particularly the charge-discharge lid seal; 2) frequent monitoring of valves, gauges, vents, and drains to see that they are operable, reliable, and used as part of the operating cycle; and 3) annual inspection by a qualified person, such as one supplied by an insurance inspection company.

Automation and Control

Batch cookers are becoming increasingly automated. The number of cycles required in a cook can be programmed. The cooker can be automatically positioned to be filled or discharged. Steam flow valves can be controlled by automation. Charts for recording time, steam pressures, and other parameters are available and helpful. Electronic controls for motion, stopping, and starting, for steam input and output, and for recording pressures, temperature, and moisture content are available from a number of manufacturers. These are becoming more numerous and efficient all the time.

CONTINUOUS COOKING

Mixing and shear in a continuous cooking process are created by the mechanical components of the cooker and resisted by the viscosity of the flowing mass of product. This interaction can generate a substantial amount of heat, which is put to use in cooking. As in batch cooking, other forms of energy are employed to cook the product as well: conducted heat and injected steam.

Energy Inputs

Inputs of energy are shown schematically in the generic continuous cooker and energy flow diagram in Figure 8. The three energy inputs are conversion (of mechanical energy from the mixing-conveying devices); conduction (of direct, externally applied heat via jacketed surfaces); and convection (of heat by the injection of steam). On the other side of the balance, energy is absorbed by the product in three ways: pump work (energy required to pressurize the product); sensible heat (energy used to raise the temperature of the product); and heats of reaction (cooking reactions are endothermic—they absorb energy to proceed). The heat absorption has been described in the literature

(Harper and Holay, 1979; Harper, 1981) by the following equation:

$$E_t/m = \int_{T_1}^{T_2} C_p \, dT + \int_{P_1}^{P_2} dP/\rho + \Delta H^0 + \Delta H_{sl} \,, \tag{3}$$

where E_t/m = total energy (E_t) absorbed per unit of mass (m), T = product temperature, C_p = specific heat of the product, ρ = density of the product at pressure P, ΔH^0 = heat absorbed by cooking reactions, ΔH_{sl} = heat of fusion of lipids, and the subscripts 1 and 2 refer to the entrance to and exit from the process, respectively.

The pump work term is usually quite small in comparison with the others and may be safely ignored, as may heat of fusion, the energy

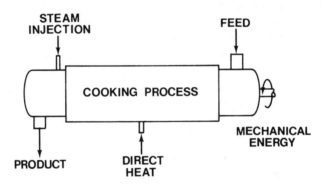

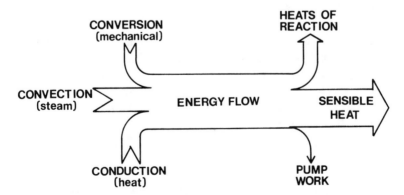

Figure 8. Generic continuous cooker and energy flow diagram.

used to melt fats in the formula—a very minor component in cereal cooking. After eliminating these terms, equation 3 reduces to the simpler form

$$E_t/m \approx \int_{T_1}^{T_2} C_p \, dT + \Delta H^0 \, , \tag{4}$$

in which the most significant terms remain: the heat of reaction and sensible heat. Heat losses also occur by conduction from equipment surfaces to the environment, but these are usually small in continuous processes characterized by small ratios of surface area to throughput. In batch systems, the heat losses can be important. In cases where the losses become significant in continuous processes, an additional term may be added to equation 4 after estimation by a standard calculation of heat transfer by air convection, as outlined in the literature (McCabe and Smith, 1956).

Heats of Reaction

The main reactions in cereal cooking are starch gelatinization and protein denaturization; minor side reactions, such as browning, also occur. The endothermic heat of reaction has been reported to be 90–100 kJ/kg for proteins (Harper, 1981) and 10–19 kJ/kg for cereal grain starches (Stevens and Elton, 1971). The heat absorbed in these reactions may be estimated by calculating the contribution of each ingredient in proportion to its formula concentration:

$$\Delta H^0 = \Delta H_s^0 + \Delta H_p^0 = \sum_{i=1}^{n} X_i \, [\Delta H_{si}^0 \, X_{si} + \Delta H_{pi}^0 \, X_{pi}] \, , \tag{5}$$

where ΔH^0 = total heat of reaction (kJ/kg of product), ΔH_s^0 = total heat of reaction for starch gelatinization, ΔH_p^0 = total heat of reaction for protein denaturization, X_i = fraction of component i in the formula, ΔH_{si}^0 = heat of reaction for starch in component i, ΔH_{pi}^0 = heat of reaction for protein in component i, X_{si} = starch fraction in component i, and X_{pi} = protein fraction in component i.

In the absence of data on the heats of reaction for particular ingredients (they do differ), the total heat of reaction can be estimated from average heats of reaction, as follows:

$$\Delta H^0 \approx 14X_s + 95X_p \, , \tag{6}$$

where X_s = starch fraction in the formula, and X_p = protein fraction in the formula.

Alternatively, one may determine the actual heat of reaction for a particular formula and process experimentally by performing an energy balance with real data and then solving equation 3 for the energy terms.

Sensible Heat

The other very significant component of energy absorption is sensible heat, which depends on the specific heat of the product as well as on the temperature rise in the cooking process. Empirical formulas for the specific heats of foods, based on their composition, have been developed (Heldman and Singh, 1981), perhaps the best for cereals being the following:

$$C_p \approx 1.424X_c + 1.549X_p + 1.675X_f + 0.837X_a + 4.187X_m , \qquad (7)$$

where C_p = specific heat of the product (kJ/[kg·K]), and X = weight fractions of constituents designated by the subscripts c (carbohydrate), p (protein), f (fat), a (ash), and m (moisture).

However, specific heat is a function of temperature as well as composition, and this becomes significant in continuous cooking, where high temperatures are often reached. A more accurate formula based on experimental data for cereal grains may be derived from equation 7 with the addition of another term for temperature (Shepherd and Bhardwaj, 1986):

$$C_p \approx 1.171[X_m(4.187 - C_d) + C_d] + 0.00304T - 0.292 \qquad (8)$$

where

$$C_d \approx 0.587X'_c + 0.712X'_p + 0.838X'_f + 0.837 ,$$

and T = temperature (°C), C_d = specific heat of the dry matter (kJ/[kg·K]), and X' = weight fraction of constituents (designated by subscripts as in equation 7) on a dry basis. This predicts values for specific heat from approximate composition, and the predicted values agree with the experimental data within about 9%, as shown in Figure 9. Separating the specific heat of dry matter is convenient for processes in which the moisture varies—a common occurrence.

From equation 8, the integrated sensible heat term in equation 4 becomes

$$\int_{T_1}^{T_2} C_p \, dT \approx \bar{C}_p(T_2 - T_1) , \qquad (9)$$

where $\bar{C}_p$ = average specific heat in the temperature range from T_1

to T_2. The average specific heat may be calculated by substituting the average temperature, $(T_1 + T_2)/2$, into equation 8.

The total energy absorbed is, of course, equal to the energy inputs in the forms of conduction, convection, and conversion:

$$E_t = E_h + E_m + E_s , \qquad (10)$$

where E_h = conducted heat, E_m = energy from mechanical conversion, and E_s = energy convected by steam injection.

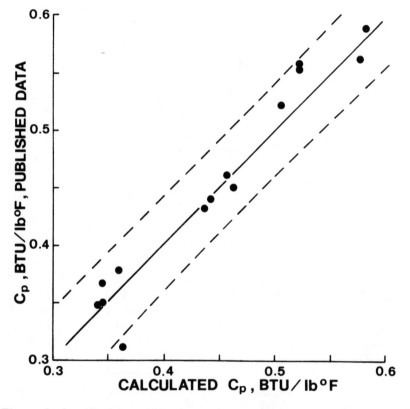

Figure 9. Specific heats (C_p) of cereals and legumes with moisture and temperature variations (0–30% moisture and 0–80°C): comparison of published data (Shepherd and Bhardwaj, 1986) and values calculated from equation 7. The English units (Btu/[lb·°F]) are numerically identical to the metric units (cal/[g·°C]). They may be converted to the SI units (J/[g·°C]) by multiplying by the conversion factor 4.184.

Energy Interchangeability

The different forms of energy are not totally interchangeable, in that they contribute to the balance in very different ways. Mechanical energy and conducted heat are both transferred to the product by the surfaces of the cooking vessel and are associated further by the heat transfer coefficient, which in turn is strongly influenced by agitation of the product (mechanical energy) at heat transfer surfaces. Without agitation, viscous materials transfer heat only by conduction within the product, which requires large temperature gradients and high temperatures at the transfer surfaces. This can cause thermal degradation of the product and nonuniform processing. These are more critical in large-scale equipment, in which the relative surface area is less than in smaller equipment—an important consideration in scale-up procedures. The rate of conductive heat transfer is given by the following equation:

$$E_h/\Delta t = UA\,\Delta T\,, \tag{11}$$

where $E_h/\Delta t$ = rate of conductive heat transfer (energy change in the time interval Δt), U = overall heat transfer coefficient, A = area of the heat transfer surface, and ΔT = difference between the temperature of the heating medium and that of the product.

Since the temperature of the product changes during the process (and that of the heating medium often changes), heat flow is not the same at all points. To calculate the overall heat transfer, equation 11 must be integrated throughout the process. The heat transfer coefficient is not constant either, since it depends strongly on the process conditions at the point of interest and on the properties of the product, both of which change as the product is cooked. However, a steady-state continuous process is always the desired operating norm. At any particular point in such a process, both heat flow and the heat transfer coefficient are constant, and average equation parameters may be determined experimentally for application to similar situations.

Typical heat transfer coefficients for various cooking methods are listed in Table 1. These are intended for rough approximation in the absence of experimental data.

Scaleup

An examination of equation 11 demonstrates a scale-up problem with conductive heating. As a process is increased in size, its volumetric capacity increases with the cube of the linear dimensions. However, the surface area used for heat transfer increases only as the square of the linear dimensions. Thus the relative area available for heat transfer decreases, and it is inversely proportional to the linear increase (i.e., a twofold increase in linear size provides only half the relative area for heat transfer). From equation 11, it is apparent that either the heat transfer coefficient, the area, or the temperature difference must be increased to compensate for this discrepancy. The simplest solution is to increase the temperature of the heating medium. This can be effective, but it puts the product in jeopardy of scorching. The heat transfer coefficient in some cases can be increased somewhat by increasing the agitation of the product at the heat transfer surfaces, but this approach is limited and can demand total changes in the process style, such as going from single-screw to twin-screw extrusion. The most common approach is to increase the heat transfer area. This can be accomplished by altering the equipment proportions (by increasing the relative length, for example) or by using other surfaces (such as those of the screw or agitator) as additional heat transfer areas.

Mechanical energy conversion is less affected by equipment size, because it takes place throughout the product mass where shear is applied and is not particularly a surface phenomenon. The uniformity of shear in the process, however, differs among cooker styles and becomes a major concern where this form of energy input predominates. Since the amount of heat generated by the conversion of mechanical energy depends on product viscosity as well as the degree of shear, it is usually most significant in low-moisture cooking.

As demonstrated by Joule in the 19th century (Schwinger, 1985), thermal, mechanical, and electrical energy are equivalent. Using this equivalence, we can calculate the heat generated by work on the product by an electrically driven mechanism as follows (1 kJ = 3,600 kW·hr):

$$E(\text{thermal}) = E(\text{mechanical}) = E(\text{electrical}) . \qquad (12)$$

TABLE 1
Overall Heat Transfer Coefficients for Continuous Cooking
(approximate ranges, w/km²)[a]

Heat Transfer Method	Starved Low-Shear Sections	Full High-Shear Sections
Steam	200–275	260–500
Water (heating or cooling)	130–240	165–385
Oils (heating or cooling)	45–210	50–320
Electric (basis: measured barrel temperature)	180–280	230–540

[a]Author Miller's data adjusted with published ranges (McCabe and Smith, 1956).

Electrical energy is best measured with a wattmeter, but it may also be calculated from line voltage and current. In the case of DC motors, this is a simple product: $W = A \times V$. For AC motors, however, the calculation is more complicated (Pumphrey, 1959):

$$\text{Power} = \sqrt{\phi}\, EI \cos \theta \qquad (13)$$

where power is the mechanical output (W), ϕ = number of AC phases, θ = phase angle (cos θ = power factor), E = line voltage (V), and I = line current (A).

Within the cooker, the transformation of work to heat is 100% efficient. At each stage of transfer before this point, however, some of the initial energy is also transformed into heat, which is lost from the process. To find the net thermal input to the product, it is necessary to reduce the original electrical energy by efficiency factors, which depend on the drive mechanism and are available from the manufacturers.

Equipment Features Affecting Formulation and Product Quality

Systems utilizing a high proportion of mechanical energy offer advantages in product formulation. A high degree of mixing permits the incorporation of a wide range of ingredients, easing the product development effort in many cases and allowing the use of nontraditional ingredients. Thus it permits fiber and protein fortification and the substitution of lower-cost ingredients that can be intimately mixed into a matrix of standard-quality materials.

An important aspect of mechanical energy is the effect of shear on product quality. Gelatinization under shear, particularly in low-moisture, high-viscosity systems, which develop high stresses within the product, can alter the starch structure through the mechanical breakdown of starch molecules, resulting in the production of dextrins and other short-chain species. In some processes, this modification of the starch is desired, but in cereal manufacturing, it can create problems. A highly disrupted starch structure generally causes stickiness that is detrimental to in-process handling of the product. Perhaps of more importance, product quality can suffer: sticky, gummy, and gooey textures and (in the extreme) dextrinized flavors can be created by overshearing the product.

Convection of energy by steam injection differs from the other inputs because it contributes not only energy but moisture to the system. Accordingly, the application of convection energy is limited to high-moisture cooking and to an open, uncompacted product structure, which permits the flow of steam through the mass. When these conditions are met, steam injection can produce a very uniform cook. The energy convected by steam injection can be expressed as follows:

$$E_s = m_s(h_s - h_w) \, , \tag{14}$$

where E_s = convective energy input (kJ), m_s = mass of injected steam (kg), h_s = enthalpy of steam (kJ/kg of steam), and h_w = enthalpy of water from condensed steam at the feed temperature (kJ/kg of steam). From standard steam tables,

$$h_s \approx 2{,}700 + 0.828 P_s \qquad h_w \approx 4.182 T_1 + 0.181 \, ,$$

where P_s = absolute steam pressure (N/cm^2), and T_1 = feed temperature (°C).

CLASSIFICATION OF COOKING PROCESSES

In practice, cereal-cooking processes use different amounts of the three available energy inputs, and they can be classified accordingly in several categories. These include *boiling-water and steam cookers; low-shear, low-pressure cookers; low-shear, high-pressure extruders; adiabatic extruders; high-shear extruders;* and *high-shear extruders with steam precookers.*

Since the energy input in these processes derives from three sources, their ranges of operation may be indicated in a triangular three-component diagram (Miller, 1985), shown in Figure 10. Every possible combination of the three inputs is a point on the triangular diagram, with the apexes representing the extreme cases of single-source forms of energy.

At one extreme, 100% conversion (the lower right apex), we find *adiabatic extruders,* which generate all of their heat through shear (Rossen and Miller, 1973). To be effective, the generation of heat entirely by conversion requires operation at low moisture (8–14%). Generally, processing moisture increases with distance from this apex, reaching a maximum of about 50% at the opposite side of the triangle, along the line representing energy sources ranging from 100% convection to 100% conduction. Points along this line represent various proportions of convection and conduction.

Boiling-water cookers use very little mechanical energy—only enough to agitate the low-viscosity mixture. Energy may be supplied by convection (injection of steam directly into the water bath, which also creates agitation, reducing the need for mechanical input) or conduction (via jackets surrounding the cooking vessel, which requires a slight amount of mechanical energy to ensure good heat transfer) or any combination of these.

Similarly, *steam cookers* use only a small amount of mechanical energy—enough to convey and agitate the product, which is loosely packed to permit steam penetration and offers little resistance. Most of the energy comes from convection, but a small amount of conduction is possible, so that the operating range of steam cooking extends somewhat into the center of the diagram from the 100% convection apex.

Low-shear, low-pressure cookers transfer energy to the product mostly by conduction from external jackets and sometimes from cored agitators designed to circulate heat transfer fluids. In these systems, mechanical energy contributes a small but significant amount of heat to the product, because shear and mixing are required for efficient and uniform heat transfer at these surfaces. Since the process does not compact the product greatly, it is possible to use some convection heating, so the

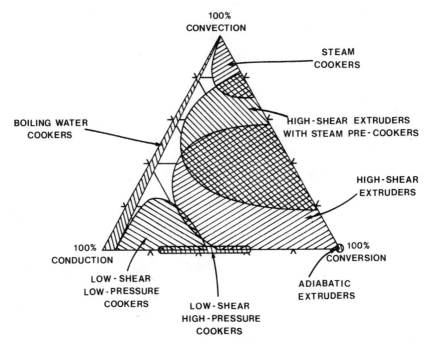

Figure 10. Three-component energy diagram for cereal-cooking processes with typical operating ranges. At any point, the energy derived from each source is proportional to its closeness to each apex. The side opposite each apex represents all combinations of the other two sources, with no contribution from the source represented by that apex.

operating range covers an area extending into the center of the diagram from the conduction-conversion line.

Low-shear, high-pressure extruders are operated at high pressure by restricting flow at the cooker exit. These cookers generate more heat by conversion of the mechanical energy needed to convey the product against the pressure gradient. They are not compatible with steam injection, because of the compacted nature of the product.

High-shear extruders are perhaps the most flexible of the available cooking processes, in that they can operate with a wide range of energy inputs. At one extreme, it is possible to run these extruders adiabatically, but a substantial amount of energy may be transferred to the product by conduction or convection. The process usually proceeds in stages, beginning with a "starved" feed section (the screw is run at a higher speed than necessary to convey the material) followed by screw segments designed specifically to provide shear or to create other starved sections where the product is susceptible to steam absorption. Conduction may be accomplished with electrical heaters, jackets, or cored screws.

Staged processing permits the performance of numerous different operations in series, with added flexibility and opportunity for optimizing any particular product. This principle is frequently further extended by the use of *high-shear cookers with continuous steam precookers*, which can increase productivity, improve product quality, and decrease operating costs. Their operating range is intermediate between that of steam cookers and that of plain high-shear extruders, and they are effective in the intermediate- to high-moisture range.

CLASSIFICATION BY PROCESSING CONDITIONS

Breakfast cereal products may be classified by the kinds of processing necessary to create their special properties, as was done in Chapter 2.

Gelatinization, like any chemical reaction, is not instantaneous; it requires time to proceed. Published experimental results (Pravisani et al, 1985) show that the reaction rate is proportional to the concentration of ungelatinized starch, which decreases exponentially with time. The proportionality constant is a function of temperature, predicted by the classical Arrhenius law (Levenspiel, 1962):

$$k = k_0 e^{-E/RT} , \tag{15}$$

where k = reaction rate constant, k_0 = frequency factor (a constant), E = activation energy (constant for a particular reaction), R = ideal gas law constant, and T = absolute temperature.

With some modification (Pravisani et al, 1985), this has been found to closely reflect the actual gelatinization rate, explaining and quantifying the accelerative effect of temperature on cooking. From these results, the time needed to reach a desired degree of cook is related to cooking temperature as follows:

$$t = (1/k_0)[\ln(X_s/X_{sf})]e^{E/RT} \propto e^{1/T} , \qquad (16)$$

where t = cooking time, X_s = initial ungelatinized starch content, and X_{sf} = final ungelatinized starch content.

The constants in equation 15 have been experimentally determined for a range of cereal grains (Pravisani et al, 1985), but in most cooking processes for breakfast cereals, the cereal formulas are too complex to allow their direct use. With experimental data on any particular formula and process, however, the forms of equations 15 and 16 are useful for interpretation and extrapolation of results and helpful in scale-up procedures.

To be of practical size, continuous cooking processes must operate at elevated temperatures, which reduce cooking time. Actual cooking methods follow this trend closely, ranging from more than 1 hr in boiling water at 100°C to only a few seconds at up to 180°C in adiabatic extruders.

Water is a reactant in gelatinization, and it also affects the process time—gelatinization proceeds more easily in the presence of excess moisture. In cereal cooking, low-moisture processes are associated with high shear stresses, however, which also increase the reaction rate by mechanically disintegrating the starch granules, countering somewhat the low-moisture effect. Water activity, not merely water content, is also important. Hygroscopic ingredients, such as salt and sugar, compete with starch for moisture, reducing the effective concentration available for gelatinization and requiring higher temperatures or longer times to achieve the same results.

The actual time needed to cook cereal products is generally a function of both moisture (interacting with shear) and temperature. With these variables, a three-dimensional graph may be constructed to show all possible cooking conditions for cereals. The typical range of conditions for each of seven cooking processes is shown as an area on the overall operating surface in Figure 11.

The operating surface is also convenient for illustrating the range of product characteristics from the cooking processes (Figure 12). These are also generally functions of time, temperature, and moisture. Low-moisture, short-time cooking requires the highest temperature and results in a dry, puffed product, typical of adiabatic extrusion.

Increasing the moisture and time decreases the temperature requirement, permitting puffing with higher moisture, typical of high-shear extruders with or without steam precookers. The high-shear cooker with a steam precooker can also produce a granular cooked product, similar to that of steam cooking, when operating under less severe conditions for a longer time. Steam cooking alone produces a granular product from a whole grain or granular feed, and boiling water is generally useful for whole grains, requiring a long cooking time at high moisture and low temperature. Low-shear, low-pressure processes, overlapping to a degree in operating conditions with the high-shear

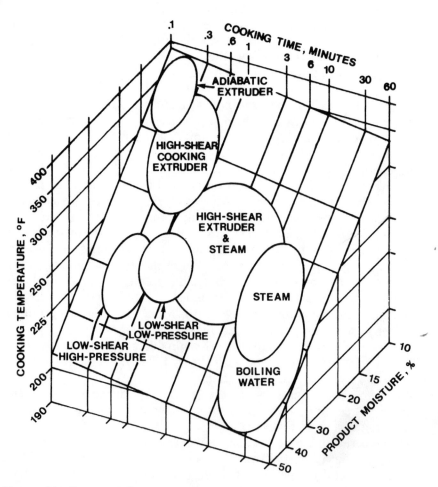

Figure 11. Range of cereal-cooking conditions—time, temperature, and moisture—with typical ranges of cooking processes.

extruders, produce a granular product. Low-shear, high-pressure extruders produce a dense, moist product, which often forms bubbles of expanded water vapor but lacks a good puffed structure.

BOILING-WATER SYSTEMS

The use of boiling water in continuous cooking systems has not become an important method, primarily because of long residence times and the associated large equipment sizes needed. The advantages of continuous operation do not easily apply to boiling-water cooking, which

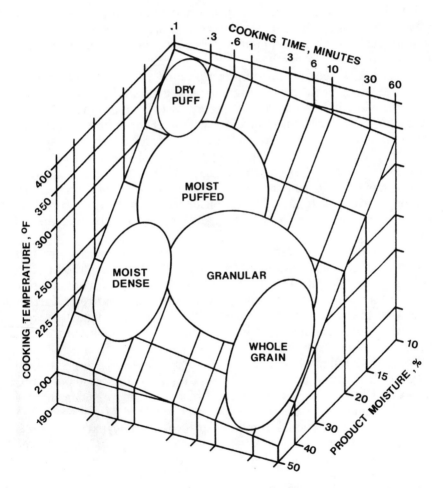

Figure 12. Product characteristics from different regions in the range of cooking process conditions.

has been used primarily for whole grain applications and where starch degradation must be absolutely minimized, as for shredded wheat.

Whole grain boiling usually requires a long holding time (at least several hours) after cooking to equilibrate product moisture within the grain. This is compatible with batch cooking, which is normally used for these products, as already described in this chapter. Nevertheless, there has been some activity (Spiel et al, 1979; Fast, 1987) in developing continuous boiling processes for special purposes.

STEAM COOKERS

Like boiling-water cooking, continuous steam cooking as a complete process is severely limited by residence time. As one component of an integrated cooking system, however, continuous steam cooking has become an important method. When it precedes high-shear extrusion cooking, steam cooking can improve productivity by adding another source of energy, thus generating more energy than the extruder alone could provide. It can also reduce energy costs by replacing expensive mechanical energy (from electricity) with a cheaper form of heat. Of more significance to the product are the effects of longer residence time and reduced shear. Replacing mechanical energy with steam generates heat with less starch damage, and the longer residence time allows moisture to penetrate the product more evenly, for a more uniform and complete cook. Typical residence times of steam precookers are from 1 min to several minutes, followed by an extrusion time of up to 1 min.

Several styles of precooker are available. Sprout-Bauer uses pressurized steam (about 10 N/cm^2, or 15 psig) to raise the product temperature to about 115°C (240°F) in a residence time of about 2 min before it is fed to the extruder, where the temperature is further increased to its maximum value by conversion of mechanical energy. A single-shafted paddle mixer is used to agitate the product for better exposure to the steam and to convey it toward the extruder.

Wenger Manufacturing uses a similar but less intense precooking step, in which atmospheric steam raises the temperature to about 100°C in about 1 min before extrusion, at which point more steam may be injected. Single- or double-shafted mixers are used in the precooker.

Anderson International combines precooking and extrusion in a system in which a large steam-injection section containing mixing blades precedes the flighted section of the screw. This design lies in a gray area between extruder steam injection and precooking. It is closer to precooking in its effect on the product, in that little mechanical energy is absorbed until the product is conveyed into the flighted part of the shaft. The residence time in the steam section is less than that of the separate units, being about 15 sec (Foley, 1979); the steam is efficiently utilized by the direct transfer to the extruder screw.

LOW-SHEAR, HIGH-PRESSURE EXTRUDERS

In the early 1950s, the idea of utilizing extrusion as a means of cooking foods began to assume significance, particularly in breakfast cereal processing. The cereal companies became pioneers in adapting equipment originally intended for ceramics or plastics manufacturing. Early successes in the field were reached in a two-stage process of cooking and forming, both done by low-shear *single-screw extruders*, a method still in common use. The process does not produce a finished product and requires a follow-up stage to shape the product. In newer systems, this is sometimes accomplished in a distinctly separate zone of the same extruder shaft.

Low-shear, high-pressure cooking uses relatively low screw speeds to generate applied shear at a shear rate of less than about 100/sec. The shear rate may be approximately calculated by applying the equation of motion (Bird et al, 1960) with the simplifying assumptions that the product is an ideal (Newtonian) fluid in an annular space between an inner, rotating cylindrical surface (the screw root) and a concentric outer, stationary surface (the barrel surface):

$$\gamma = 4\pi N (r_i R)^2 / r^2 (R^2 - r_i^2) , \tag{17}$$

where γ = shear rate (1/time) at radius r, N = screw speed (in revolutions per unit of time), r_i = inner annular radius (screw root), and R = outer annular radius (screw tip or barrel surface).

The average shear may be found by integrating this function over the cross section of the annular volume of the material under shear:

$$\bar{\gamma} = 8\pi N [\ln(R/r_i)] (R/r_i)^2 / (R^2 - r_i^2)^2 , \tag{18}$$

where $\bar{\gamma}$ = average shear rate between r_i and R. In the case of a thin annulus (shallow screw flight), this may be approximated by the simpler form

$$\bar{\gamma} = \pi N (R + r_i) / (R - r_i) . \tag{19}$$

These expressions may be used to estimate the product shear at any point along the extruder length for rough comparisons. For a more rigorous calculation, the actual rheological properties of the product must be taken into account. These are definitely not Newtonian for foods and are a subject of active current research (Clark, 1982). The power law model, in which the apparent viscosity is inversely proportional to the applied shear rate raised to a characteristic power— for cereals it decreases with shear—has been used to model flow behavior (Jao et al, 1978). The situation is further complicated by other flow components not included in the concentric-cylinder model. These have been analyzed for simple systems, leading to calculation of a "weighted average total strain" throughout the extruder (Tadmor and Klein, 1970). Since these models apply only to simple geometries, they are of limited use for food-cooking extruders, which frequently use complex empirical designs.

The feed material is usually a dough or wet granular mix (about 30% moisture), which is choke-fed; this means that the feed hopper is kept full and the feed rate is determined by the ability of the screw to draw material from the hopper. The Bonnot system uses a long screw and barrel (the ratio of screw length to diameter is about 20) to generate extrusion pressure and to provide sufficient surface area for efficient heat transfer from the surrounding steam jackets. Most of the cooking energy is supplied by this conductive heating, but a substantial amount (about 30%) is generated by conversion of mechanical energy, which is required for mixing, good heat transfer, and homogenization of the product. This mechanical energy is created by high-pressure operation (several hundred newtons per square centimeter or pounds per square inch) caused by discharging the product through a small orifice, or die.

At the normal high-moisture levels, this mode of operation creates a large backward "pressure flow" component. The Bonnot type of extruder also normally utilizes a compression screw (up to a 9:1 reduction in channel volume) with variations in both screw pitch and channel depth. This is useful in eliminating air in the product by squeezing it out of the feed hopper, and it is even more important in generating the desired pressure profile, which typically reaches a maximum before the product reaches the die. Grooved barrels are also the norm. These serve two functions: reduction of slippage at the barrel surface and increase of back-flow over the flight tips, where shear and heat transfer are maximum.

Another type of low-shear extruder, also borrowed from the plastics industry, has been adapted for cereal cooking—the *counterrotating twin-screw extruder* (see Figure 5 in Chapter 6). As described in more detail in Chapter 6, the principle of this extruder differs in several important ways from that of the single-screw. It is more of a positive-displacement pump and therefore does not generate much heat by conversion of mechanical energy. The flow patterns developed in the intersecting screw channels create sufficient mixing for good heat transfer, so that conduction heating is efficient. Enormous pressures can be generated if desired, even with low-viscosity materials, and this capability is particularly useful for high-moisture cooking. If shear is desired to develop product characteristics, it must be generated in a separate process. Textruder accomplishes this with their tapered- and straight-screw models by supplying a follow-up unit consisting of a single-screw extrusion stage or a special shear disk unit. Similar equipment is available from Cincinnati Milacron. Twin-screw extruders of either the counterrotating or the corotating variety (discussed next) may also be used for high-shear cooking.

A great deal of attention is now being paid to the latest style of extrusion cooker, the *corotating intermeshing twin-screw extruder*, also adapted from the plastics industry (see Figure 10 in Chapter 6). It has been widely used for foods only since about 1980. Several companies offer similar equipment, including Clextral, which made the earliest significant penetration into the food industry, followed quickly by Werner & Pfleiderer and APV Baker, which had also been suppliers of plastics machinery. Responding to competitive pressure, companies producing single-screw food extruders, including Wenger and Buhler, have introduced their own versions of the corotating twin-screw.

The corotating twin is the most versatile (and expensive) style of extruder available, in that it can operate over a range of conditions including those of most of the other, more specialized machines, depending on the particular screws and operating variables selected. A self-wiping action between the intersecting screws makes it very effective as a heat exchanger, so that conductive heating in the low-shear mode can be used efficiently. In this mode, where screws are selected to limit shear, the residence time distribution is quite narrow (with the result that the product receives more uniform processing), although it is still wider than that of the counterrotating twin. It is difficult to differentiate between high- and low-shear operation in the corotating twin except by an arbitrary standard, since there is a continuum of possible conditions as more shear is added with reverse-pitch screws and mixing elements and with increasing screw speed, which is variable over a wide range up to about 400 rpm.

As a preforming high-moisture cooker, however, the corotating machine does the same job as the low-shear, high-pressure cooking extruder. It requires a postcooking step, which can be performed by a single- or twin-screw extruder, a forming stage within the cooker, or some other forming step for the cooked, plastic (high-moisture) material.

LOW-SHEAR, LOW-PRESSURE COOKERS

Several kinds of equipment have been developed for continuous cooking under low-shear, low-pressure conditions for minimum starch degradation. This is a category with no common underlying principles beyond the main goal. The following examples illustrate some of the approaches that satisfy the stated needs.

Mapimpianti has developed a cooking process based on an extended macaroni extruder design. It uses a single-screw extruder with a series of compression stages that redistribute the flow for improved heat transfer in the absence of high shear or pressure development. The product is not compacted and remains in a granular state throughout the process, which makes it suitable for the incorporation of other ingredients after cooking and before forming. A modular process line includes an intermediate mixing-tempering stage. To maintain the desired granular structure, the product is not forced through a die; it emerges through an essentially open discharge after the last compression stage. Small, stationary blades at the discharge are used to break up agglomerates.

A different approach is used in Teledyne's Readco cooker, which is based on the principle of the corotating twin-screw extruder, but with exaggerated geometrical proportions. Since this cooker was developed as a continuous mixer, the screw flights are very deep and taper at their tips to become very thin in cross section. Efficient mixing is achieved by the transfer of material between the screws. This does not consume a great deal of mechanical energy in the Readco configuration, but it is conducive to good heat transfer from the jacketed barrel and, in the larger sizes, from cored screws. The machine is not capable of generating high pressure, and it discharges through a large opening, which is usually fitted with a swinging gate to control resistance. This in turn controls residence time and mechanical energy consumption. These are important factors in determining the degree of cook. Alternatively, the cooker can discharge directly into a single-screw extruder, mounted perpendicularly to the cooker axis, to form the product or generate pressure when desired.

A third approach to low-pressure cooking is taken in the unique Buss-Condux design, which has a similar origin in continuous mixing. This design uses a single cut-flight screw that intermeshes with blade-shaped stators extending inward from the barrel. The screw not only rotates but reciprocates axially, so that the stators are wiped on both the upstream and the downstream surfaces by the screw segments, which in turn are also wiped. In many respects, this mechanism accomplishes the same ends as the corotating twin-screw extruder, with similar advantages in heat transfer efficiency. Like the Readco machine, however, it cannot generate high pressure and must be linked to other processes to complete the system.

ADIABATIC EXTRUDERS

Adiabatic extrusion combines cooking, forming, puffing, and partial drying in one high-temperature, short-time process. It is a specialized method limited to direct-expanded products, discussed separately in Chapter 6.

HIGH-SHEAR EXTRUDERS

High-shear extrusion is the most versatile of the cooking processes. Wide ranges of operating conditions and product characteristics are made possible by the various options for energy input. Commercial extruders, in single- and twin-screw styles, have some or all of the features shown in the generalized drawing in Figure 13.

Density

Cereal ingredients in an uncompacted state have a density of about 0.43–0.80 g/cm^3 (Carr, 1976). Unless they are made into a high-moisture dough with fluid properties, this low density is maintained in feeding devices where the material must be capable of flowing easily. Once it is in the cooking process, the feed is subjected to pressure and shear, which compact it to a density of around 1.2–1.6 g/cm^3, depending on moisture and formulation (Peleg and Bagley, 1983; Malave et al, 1985).

To accommodate the low feed density, feed screws must be of long pitch, with high volumetric capacity. The situation in twin-screw machines is aggravated by their conveying efficiency in the feed throat. In fully developed flow, material shifts back and forth between the intermeshing screws. This pattern takes time to reach equilibrium, however, so that in the feed section, one screw—the one that pulls the feed downward—runs full, while the other is more or less empty for the time required for product transfer between screws. This results in a volumetric efficiency as low as 50% for the first increment of screw length in the feed section.

Since the feed material in high-shear cooking is usually granular, the feed section is normally run starved, to permit easy flow into the extruder throat. In this situation, the feed rate is externally controlled and maintained at a point below the volumetric screw capacity, further reducing conveying efficiency.

Mixing

Once past the feed throat, at a point where flow patterns have been developed within the surrounding barrel, the screw pitch is reduced, to start compressing and intensively mixing the feed. In some processes,

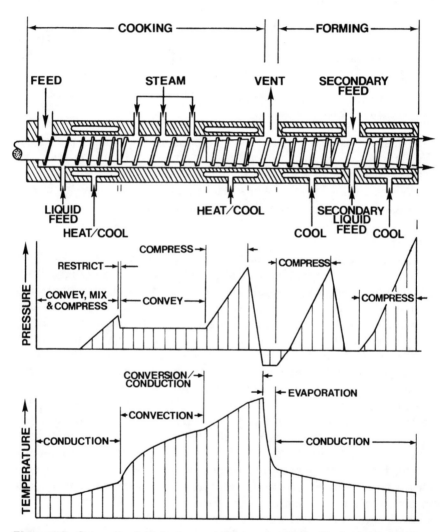

Figure 13. Generalized high-shear cooking extruder with integral forming section (top). Schematic pressure profile corresponding to sections of the high-shear extruder, with mechanical functions of the screw segments (center). Schematic temperature profile of the extruder, with zones of energy input (bottom).

special mixing elements are used for this purpose. They may consist of cut-flight or reverse-pitch screws or, in the twin-screw extruder, lobe- or disk-shaped kneaders. The mixing section is useful for intimate blending of solids, which may be fed as separate streams into the feed section, and for the incorporation of liquid streams, which are commonly injected through the barrel at this point. This configuration is sometimes repeated at other points along the barrel in staged processes that require the addition of other ingredients after some processing has been completed—a useful way of adding heat-sensitive components, such as flavorings.

Heating or Cooling

Conductive heating or cooling may be employed in the feed and mixing sections. Cooling can help to develop flow patterns for pressure generation and can condense any water vapor flowing back from the high-temperature cooking section. Backward vapor flow is a particular problem in single-screw extrusion, in which steam rising through the feed throat can interfere with feeding and create a mess. Where these problems do not exist, the cooking process may be given a head start by adding some heat in the mixing zone.

The screw channel is commonly full at the mixing zone exit, creating a barrier to backward vapor flow. Where desired, a full channel is purposely created by the selection of screws, mixing elements, or flow restrictors to cause the material to build up as the pressure rises. With a barrier in place, steam may be injected through the barrel to accomplish the first cooking stage. (Steam injection is done in a separate operation in some systems, as noted under Steam Cookers, and not at all in others.) The degree of heating possible with steam cooking is limited by the allowable moisture gain in the product (for low-moisture extrusion it is not appropriate) and by practical limitations on steam pressure (for continuous cooking with a short residence time, at a temperature of 180°C [350°F], a steam pressure of at least 100 N/cm^2 [150 psig] would be required). As a first stage, however, it is an efficient way of increasing product temperature rapidly and uniformly without the danger of scorching (steam "seeks out" and condenses on the coolest portions of the product), and it is not subject to the scale-up limitations of heat exchange surfaces. To ensure intimate and effective contact between the steam and the product, however, a sufficient open volume must be provided within the barrel. This is done by creating another starved section by means of long-pitch screws.

Cooking

The final cooking stage is done at elevated pressure and under high shear, which serve two functions: generation of heat by conversion of mechanical energy and improvement of heat transfer efficiency. In the final cooking stage, conductive heating from a jacketed barrel (steam or heat transfer fluids) or electric heating elements (resistance or induction) may be applied, but it is limited by heat transfer rates. The final temperature rise is therefore normally accomplished by conversion of mechanical energy, discussed in more detail in Chapter 6.

Pressure in the final cooking stage is maintained by forcing the product to exit through a restrictive orifice. If the product emerges from the process in its superheated state, it rapidly flashes off enough moisture to cool it down to equilibrium with the atmosphere. This effect may be increased by flashing the product off into a vacuum, accomplished by the use of a vent port through the barrel wall. Venting requires that the fluid pressure be reduced to zero, to prevent flow into the vent. This is ensured by creating another starved section with long-pitch screws, and it is more easily done with twin-screw extruders, which have better pumping characteristics. In cases where rapid cooling is not needed and product expansion is to be avoided, conductive cooling may be applied after cooking.

Barrel vents may also be used for feeding additional ingredients to the system after cooking is completed. This secondary feed is especially advantageous for the addition of heat-sensitive materials. It is normally used only when followed by an integral cooled forming section, as discussed in Chapter 6.

Lump Breaking and Sizing

Lump-breaking and sizing or screening operations in cereal processes are two separate steps, usually in close proximity to each other. Multiple pieces of equipment are often used for both operations, since they are frequently performed in a stepwise fashion. Coarse breaking takes place first, followed by coarse screening, then finer crushing, followed by finer screening, and so on until the product reaches the desired final size. Performing these operations in a stepwise manner minimizes the production of unwanted fines.

Lump-breaking and sizing operations are described in this chapter after the discussions of cooking because most often when these two steps are required in a breakfast cereal process, they are required after cooking. Masses of cooked grain when dumped from a batch cooker are usually just that—nondescript masses much bigger than the desired finished cereal pieces. Lump breaking is essential for obtaining smaller pieces of uniform size that can be dried, tempered, and processed into the final cereal shape.

LUMP BREAKING

Since lump breaking is usually needed just after material is dumped from a batch cooker, it is performed on a hot, sticky mass of cooked grain and other ingredients. Jam-ups easily occur in batches that have been cooked out of control, i.e., batches that have been overcooked or have a higher moisture content than the process calls for. In cases such as this the breaker then turns into a powerful mixer that destroys the product rather than performing its breaking function.

One feature usually employed in most breaking and sizing operations is the use of large volumes of air drawn through the equipment. In

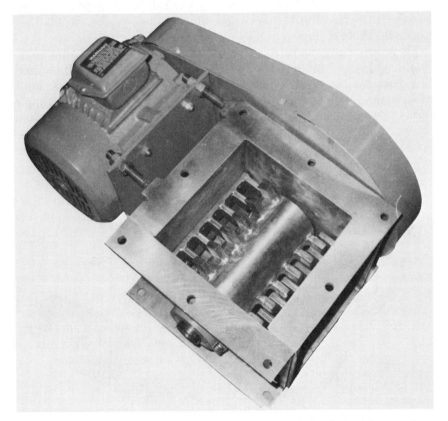

Figure 14. Lump-breaking machine (Lump Abrador), designed to reduce agglomerated material that is friable and can be reduced in size by the impact of rotating crushing and impact fingers mounted on a rotary drum. The impact fingers pass through two sets of combs. Products up to 1 in. in size can be reduced to particles as fine as 30-mesh. (Courtesy Jersey Stainless, Inc., Berkeley Heights, NJ)

the breaking section this helps to cool the product, slightly skin it over and thereby reduce its stickiness, and remove steam vapor, which is released as the cooling takes place. In the sizing section the air also removes heat and moisture vapor and fine dust or particles from the product stream.

Another aspect of proper breaking and sizing is the uniformity of the product feed to the units. It is not usually necessary to employ gravimetric feeders; well-measured volumetric feeding is sufficient. This is most easily accomplished by controlling the speed of the conveying belt ahead of the breaker or sifter. Speed control is also used in combination with adjustable gates mounted above the feed conveyors, which produce a uniformly thick layer of product on the feed belt. Overfeeding breakers and sifters is one of the pitfalls that can cause them to jam.

There are many manufacturers of these units, including Champion Products, Jersey Stainless Inc., Gruendler Co., Franklin P. Miller, and S. Howes Inc., to name a few. Most breakers are rugged units, built to withstand the pounding incurred in the breaking process. In simple terms they consist of the following parts: 1) the main body, often equipped with a fixed comb or fingers, 2) one or more rotating shafts, on which breaking bars or projections are mounted, 3) a grid or screen

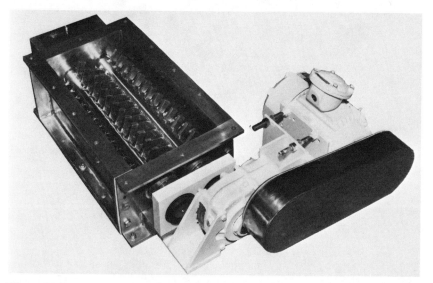

Figure 15. Lump-crushing machine. Dual counterrotating shafts are fitted with crushing fingers or bars. The machine can be fitted with special screens or plates to hold the product back until it reaches the desired particle size. (Courtesy Jersey Stainless, Inc., Berkeley Heights, NJ)

device to hold the product in the breaker until the desired particle size is achieved, and 4) a drive train, consisting of a motor with or without a speed reducer.

Representative types of lump breakers are shown in Figures 14–17. Each unit is slightly different in design and construction. As a result they handle different products with varying degrees of efficiency in breaking, generation of fines, and jamming tendencies (Anonymous, 1985b; Feldman, 1987). The units are typically offered in carbon or stainless steel, with the latter preferred for food use.

SIZING

Like lump breakers, sifters come in various sizes and shapes, each with its own unique application properties. Some manufacturers are Rotex, Inc. (Anonymous, undated), Azo Inc. (Anonymous, 1985a), and Sweco Inc. The Rotex and Sweco machines are vibrator and gyrating types, the former being rectangular single- or multidecked and the latter being round single- or multidecked. Azo screeners are different in that

Figure 16. Heavy-duty crusher (Titan). This machine performs high-torque crushing at low speed to minimize the generation of fines, with split-grid construction for product holdback. The grids may be removed through the sides of the unit, obviating the need to remove the whole unit from the line. (Courtesy Champion Products, Inc., Eden Prairie, MN)

the screen in these units is cylindrical, horizontally mounted, and stationary. A rotating paddle assembly is mounted inside the screen, and it is this that feeds the product through the screen, from the inside to the outside.

Figures 18–20 show how a Rotex screener is built. In Figure 19, the uppermost deck, on which the mixed-size material first alights, is the coarser of the two deck screens. The lower deck separates out midsize particles as overs and throughs. The circles under each screen

Figure 17. Medium-duty crusher (Gladiator). Pins mounted in the main body come in different combinations for different operations and are removable from outside the unit. The shafts are rotated toward each other for delumping and away from each other for granulation. The actual particle size depends on the holdback screen used. (Courtesy Champion Products, Inc., Eden Prairie, MN)

deck in the sifter represent resilient balls, which prevent screen blinding.

Figure 20 depicts the path the material takes in flowing from the infeed to the discharge end of the sifter. This is represented by the circular black lines and arrows. The gyratory motion performs three functions in this type of sifter. First, it spreads the material across the full width of the screen deck. Second, it stratifies the material, causing the fines to sink down against the screen, where they quickly pass through the screen openings. Third, it allows the larger particles to float to the top, where they are conveyed toward the discharge. The bouncing balls underneath the screen are deflected against bevel strips and bounce continuously against the underside of the screen mesh. This action continuously cleans the screen mesh openings.

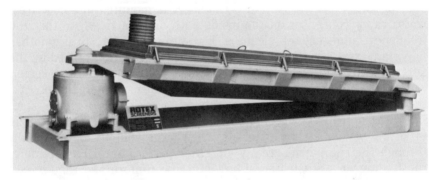

Figure 18. Gyratory sifter (Rotex Screener). (Courtesy Rotex, Inc., Cincinnati, OH)

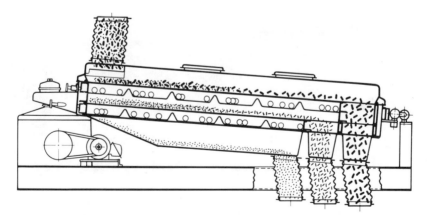

Figure 19. Vertical cross section of Rotex Screener.

The Azo Cyclone screener consists of a horizontally mounted cylinder of nylon screen cloth on a metal frame. A short motor-driven screw conveys the feed stock into the screening cylinder, where the whirling paddles, mounted on the same shaft as the feed screw, force the material through the screen opening. The natural vibration of the screen cloth performs the cleaning action. The paddles also have a lump-breaking action on any soft, friable lumps in the feed stock.

All of the units as shown or described are totally enclosed and relatively dust-tight in operation. Some of the questions that have to be considered in choosing a unit are as follows. The first consideration is matching the flow rate to the required square footage of the screen area, so that particle size ranges of each fraction from the screener meet the necessary specifications. Another consideration is the temperature of the infeed material—is it warm enough to be giving off moisture vapor? If so, and if the sifter is mounted in a cool area, water vapor condenses on the inside with dust particles, so that the frequency and ease of cleaning become important. Still another important consideration is possible product degradation caused by the sifter. Sometimes the rubbing action of the product against the screen causes attrition or product breakdown. In some cases where this occurs, the product itself may have to be altered to withstand a sifting step.

As was noted in the section on lump breaking, there are occasions in sifting wet or sticky materials when large volumes of air drawn through the equipment greatly aid sifter performance. Furthermore, the feeding of stock to the sifters must be uniform, so that screen areas do not become overloaded. Overloaded screen surfaces fail to make the desired separations; fines cannot pass through an overloaded screen and are carried over with the overs stock.

Figure 20. Rotex Screener from above, with the cover removed. The lines and arrows indicate the path of material from the infeed to the discharge. (Courtesy Rotex, Inc., Cincinnati, OH)

The importance of proper lump breaking and sizing cannot be overstressed. The size of the finished product and its ultimate package weight may be a function of how well the lump breaking and sizing were carried out earlier in the process. This is particularly true of some flaked cereals, some granolas, and several hot cereal products, particularly instant mix-in-the-bowl hot cereals.

References

ASME. 1989. ASME Boiler and Pressure Vessel Code, Sec. VIII, Div. 1, para. UG 35. American Society of Mechanical Engineers, New York.

Anonymous. 1985a. Cyclone Screener Bull. 7/85-5M. Azo, Inc., Memphis, TN.

Anonymous. 1985b. Bull. GLC 85.4 and Bull. TIC 85.4. Champion Products, Inc., Eden Prairie, MN.

Anonymous. 1981. Catalog 600-81, I. S. Howes Co., Inc., Silver Creek, NY.

Anonymous. n.d. Catalog 806. Rotex, Inc., Cincinnati, OH.

Bird, R. B., Stewart, W. E., and Lightfoot, E. L. 1960. Transport Phenomena. John Wiley & Sons, Inc., New York.

Björck, I., and Asp, N.-G. 1984. The effects of extrusion cooking on nutritional value—A literature review. Pages 181-208 in: Extrusion Cooking Technology. R. Jowitt, ed. Elsevier Applied Science Publishers, New York.

Bruin, S., Van Zuilichem, D. J., and Stolp, W. 1987. A review of fundamental and engineering aspects of extrusion of biopolymers in a single-screw extruder. J. Food Process Eng. 2:1-37.

Carr, R. L. 1976. Powder and granule properties and mechanics. In: Gas-Solids Handling in the Processing Industries. J. M. Mardhello and Gomezplata, eds. Marcel Dekker, Inc., New York.

Clark, J. P. 1982. Rheology in food extrusion. AIChE Symp. Ser. 78:218.

Davidson, V. J., Paton, D., Diosady, L. L., and Spratt, W. A. 1983. Residence time distributions for wheat starch in a single screw extruder. J. Food Sci. 48(4):1157-1161.

Fast, R. B. 1987. Continuous process for cooking cereal grains. U.S. patent 4,699,797.

Feldman, H. 1987. Selecting a lump breaker for gross size reduction. Powder Bulk Eng. 1(6):26-30.

Foley, K. M. 1979. The optimum timing of the addition of moisture in a screw extrusion process—A case history with pet food. In: Proc. Inst. Briquet. Agglom. Bienn. Conf., 16th, San Diego, CA.

Harper, J. M. 1981. Extrusion of Foods. CRC Press, Boca Raton, FL.

Harper, J. M., and Holay, S. H. 1979. Optimal energy usage in food extrusion. ASAE Pap. 79-6508.

Heldman, D. R., and Singh, R. P. 1981. Food Process Engineering. AVI Publishing Co., Westport, CT.

Janssen, L. P. B. M. 1978. Twin Screw Extrusion. Elsevier Scientific Publishing Co., New York.

Jao, Y. C., Chen, A. H., Lewandowski, D., and Irwin, W. E. 1978. Engineering analysis of soy dough rheology in extrusion. J. Food Process Eng. 2:97-112.

Levenspiel, O. 1962. Chemical Reaction Engineering. John Wiley & Sons, New York.

Malave, J., Barbosa-Canovas, G. V., and Peleg, M. 1985. Comparison of the compaction characteristics of selected food powders by vibration, tapping and mechanical compression. J. Food Sci. 50:1476.

McCabe, W. L., and Smith, J. C. 1956. Unit Operations of Chemical Engineering. McGraw-Hill, New York.

Miller, R. C. 1985. Extrusion cooking of pet foods. Cereal Foods World 30:323.

Peleg, M., and Bagley, E. B. 1983. Physical Properties of Foods. AVI Publishing Co., Westport, CT.

Pravisani, C. I., Califano, A. N., and Calvelo, A. 1985. Kinetics of starch gelatinization in potato. J. Food Sci. 50:657.

Pumphrey, F. H. 1959. Fundamentals of Electrical Engineering. Prentice-Hall, Englewood Cliffs, NJ.

Rossen, J. L., and Miller, R. C. 1973. Food extrusion. Food Technol. 27(8):46-53.

Schwinger, J. 1985. Einstein's Legacy. Scientific American Books, New York.

Shepherd, H., and Bhardwaj, R. K. 1986. Thermal properties of pigeon pea. Cereal Foods World 31:466.

Spiel, A., Kim, S. K., Schutt, S. H., and Arthur, J. 1979. Continuous cooking apparatus and process. U.S. patent 4,155,293.

Stevens, D. J., and Elton, G. A. H. 1971. Thermal properties of the starch/water system. I. Measurement of heat of gelatinization by differential scanning calorimetry. Starch/Staerke 23(1):8.

Tadmor, Z., and Klein, I. 1970. Engineering Principles of Plasticating Extrusion. Van Nostrand Reinhold Co., New York.

Tressler, D. K., and Sultan, W. J. 1975. Food Products Formulary, Vol. 2: Cereals, Baked Goods, Dairy and Egg Products. AVI Publishing Co., Westport, CT.

Williams, M. A., Horn, R. E., and Rugala, R. P. 1977. Extrusion: An in-depth look at a versatile process. Food Eng. 49(9):99.

Chapter 4

Unit Operations and Equipment II. Drying and Dryers

BRITTON D. F. MILLER

Basic Principles of Drying

As pointed out in Chapter 2, most ready-to-eat cereals require drying as an intermediate processing step. This drying is the controlled removal of water from the cooked grain and other ingredients to obtain appropriate physical properties for further processing such as flaking, puffing, forming, toasting, or packaging. The desired properties might include formability, internal viscosity, flavor, and shelf stability as well as reduced moisture content.

Cereals may be dried at several points. The cooked cereal mass may be "predried" to prevent agglomeration and product damage in further material handling and to create appropriate material properties before flaking or puffing. Because drying is used to change material properties for specific processing reasons, the existence of a moisture gradient in the mass as it leaves the dryer may need to be alleviated by tempering to create a more uniform moisture content within and among the cereal particles. Puffing and toasting operations involve further drying of products as well as changing their physical structure and chemical makeup. After a coating (e.g., a sugar syrup) has been applied, cereals are dried to set and harden the coating and to remove excess moisture.

Predrying and the dryers used for predrying and sugar coating are examined in this chapter. Puffing and toasting operations and the associated equipment are considered in Chapter 5, whereas this chapter focuses mainly on the cooked cereal mass and on cereal pellet drying and dryers. Important process and product parameters are defined and discussed. Finally, the application of the basic principles and parameters of drying to the design and selection of the dryer best suited to particular unit operations is addressed.

Only a limited treatment of the principles and theories of drying is included here. The theory is well developed and the literature extensive. Most texts on drying have bibliographies for further reference (e.g., Sherwood, 1929; Slesser and Cleland, 1962; Williams-Gardner, 1971). Accordingly, the mathematical and physical derivations and details available elsewhere will not be repeated here.

The thermal drying process is the result of simultaneous heat and mass transfer whereby water is vaporized and removed from the product. In the case of particles or pellets of cooked cereal, the dominant mechanism of heat transfer is convection from air to the product and water, and the dominant mechanism of mass transfer is diffusion after the surface moisture has evaporated.

In the discussion that follows, constant drying temperature and relative humidity are assumed to prevail inside the dryer.

DRYING RATE

Initial Constant-Rate Period

For a short time, pellets behave as if the surface in contact with the air is completely wetted. The drying rate is constant and is determined by the rate of water evaporation, which depends on surface water temperature; air temperature, humidity, pressure, and velocity; and even to some extent the size and shape of the surface and the direction of air movement (Van Arsdel et al, 1973). In actuality, the surface may not be completely wetted but may have "pools" on a porous or concave structure; or the water within the piece may move fast enough that the surface behaves as if it were wet. In either case, the drying rate is very nearly constant and is assumed to be driven by the evaporation rate.

Idealized drying curves are depicted in Figure 1 on both a moisture content and a drying rate basis. For a short time, corresponding to line AB in Figure 1, the product and water in it are being heated to the temperature at which water evaporates most readily. The constant drying rate period corresponds to line BC; the rate of drying is essentially the evaporation rate, as if no solid were present. The temperature of the product surface and of the water during this period is close to the wet bulb temperature of the air. The factors of most importance in determining the drying rate are the heat and mass transfer rates, the surface area exposed to airflow, and the differences between the temperature and humidity of the air and those of the wet product surface

(Williams-Gardner, 1971). The heat and mass transfer rates are governed by the associated transfer coefficients, which in turn are directly related to the velocity and temperature of the air over the product; that is, an increase in air velocity and/or temperature will increase the drying rate. The driving force for evaporation is the difference between the water vapor pressure of the water in the product and the partial pressure of water vapor in the airflow impinging on the product. Any increase in the difference between the dry bulb and wet bulb temperatures of the air also accelerates drying.

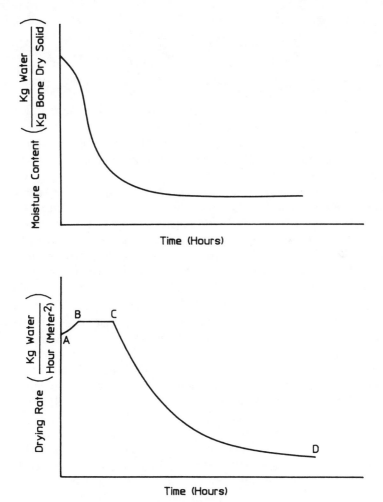

Figure 1. Idealized drying curves. Top, moisture content versus time; bottom, rate of drying versus time.

There are basic limitations to increasing the drying rate during the constant-rate period. Air velocity is limited by the incipient fluidization of the particle for flow up through a bed and by bed compaction for downflow. Products may undergo chemical or physical degradation such as caramelization and burning if the temperature is too high, although to the extent that this occurs, the drying rate is no longer constant. Finally, there is a trade-off between maximum air humidity driving force and energy considerations, which is discussed further later in this chapter.

Falling-Rate Period

At a certain moisture content, all the surface moisture (or moisture that can be transported to the surface easily and quickly) has evaporated and the surface is no longer completely wetted. The drying rate slows, as shown by curve CD in Figure 1. This is called the falling-rate period. Some products may show multiple falling-rate periods. Most drying of cereal pellets falls into this range.

The moisture content of the product at the break point on the drying curve (point C) is called the *critical moisture content*. The critical moisture is not a specific food property but is related to the physical characteristics of the pellet (such as size, shape, structure, and density) and the properties of the surrounding air (such as temperature, humidity, and velocity) (Karel, 1975). Because the material only appears to behave as if the surface were wetted, which may not actually be the case, a specific critical moisture content and a sharp delineation between the constant- and falling-rate periods may not be clearly observed.

In the falling-rate period, drying occurs by the evaporation of water from a receding front where the assumed conditions are those of saturation. The moisture is assumed to be vaporized and then carried in the gas phase to the surface to be swept away in the air. The dominant mechanism affecting the drying rate is the diffusion of the moisture to the surface. Sophisticated mathematical treatment and discussion of this phenomenon can be found elsewhere (e.g., Crank, 1956). The actual mechanisms may not be diffusional at all and on the molecular level may involve adsorption and bonding forces as well as the more macroscopic difficulties of cell structure (Van Arsdel et al, 1973). However, the mathematical theory of diffusion provides a practical model for the observed drying behavior of cereal pellets. Wilgen (1979) presents an excellent model and discussion.

The governing differential equation for drying of spherical pellets is:

$$\frac{\partial M}{\partial t} = \frac{D}{r^2} \frac{\partial}{\partial r} \left(\frac{r^2 \partial M}{\partial r} \right),$$ (1)

where M is moisture content (dry basis), D is the diffusion coefficient, t is time, and r is the local body radius. Using appropriate boundary conditions and computing the average moisture content with the volume integral, the standard solution is:

$$\frac{M - M_e}{M_o - M_e} = \sum_{n=1}^{\infty} \frac{6}{(n\pi)^2} \exp \left(\frac{-n^2 \pi^2 Dt}{R^2} \right),$$ (2)

where M_o and M_e are the initial and equilibrium moisture contents, respectively, and R is the pellet radius. Most drying curves of starch-based cereal pellets can be very adequately characterized by equation 2.

In this model, moisture migration is described in terms of the diffusivity of water and the moisture gradient in the pellet (Van Arsdel et al, 1973). However, in diffusion theory, the diffusivity is normally assumed to be constant (see the form of equation 1). Crank (1958) and others have developed procedures to account for this nonconstant diffusivity. Wilgen (1979) used a computer fitting routine to determine the following empirical relationship for the diffusion coefficient:

$$D = 8.68 \exp(2.96 \text{ RH})\exp(-5,580/T),$$ (3)

where RH is the relative humidity of the air and T is the absolute temperature in degrees Kelvin.

Drying in the falling-rate period thus depends mainly on air temperature, air relative humidity, and particle thickness, if the airflow rate past the product surface is assumed to be enough for adequate heat transfer. That the rate of drying varies as the square of the particle thickness is easily derived from equation 2. This assumes that only the first term is important, which is the case for calculating the ending moisture content of pellets.

The most noticeable change in drying rate is accomplished by increasing the dry bulb temperature of the air, which decreases the amount of energy needed for vaporization and also increases the diffusivity of water inside the particle. Van Arsdel et al (1973) suggested that the temperature increase might also increase diffusion by a water vapor pressure gradient.

Experimental data (Wilgen, 1979) have shown that increasing the relative humidity of the air also increases the drying rate. It may also be deduced that an increase in air relative humidity raises the temperature of the particle at the evaporation front, thereby increasing drying (Williams-Gardner, 1971).

Wilgen (1979) also discussed how the relative humidity of the drying climate affects case hardening. A dry climate may dry the outside of the particle rapidly, changing its physical and chemical structure and rendering it less permeable, if not impervious, to water vapor. While it is obvious from the perspective of temperature driving force that raising the relative humidity slows the drying rate, the changes in drying that result from case hardening are more significant for the particular conditions of pellet drying. If the case is formed, drying is quite inhibited. An increase in humidity will slow or significantly decrease case formation. It also increases diffusivity (see equation 3) and the drying rate.

DRYING PROCESS PARAMETERS

The most important drying parameters are the basic physical properties of the cereal product and the properties of the air used as the drying medium. A brief discussion of the moisture properties of air is needed to provide a basis for understanding the important air parameters. Details may be found elsewhere (e.g., ASHRAE, 1977).

As mentioned previously, drying involves simultaneous heat and mass transfer: heat transfer to raise the temperature of the solid and its associated moisture and to evaporate the moisture; and mass transfer to carry the water from the evaporation surface to the particle surface and away into the air. The amount of air needed to carry away the moisture, as distinct from the amount of air needed to transfer enough heat to cause evaporation, is determined by the ability of the air to pick up additional moisture (Williams-Gardner, 1971).

The absolute humidity of the air is the number of kilograms of water associated with a kilogram of dry air. The humidity of the air (W) depends on the partial pressure of water vapor in the air (P_w) and the total system pressure (P_t) as follows:

$$W = \frac{M_w}{M_a} \left(\frac{P_w}{P_t - P_w} \right), \tag{4}$$

where M_w and M_a are the molecular weights of water and air, respectively. If the sample of air is cooled, it becomes saturated (i.e., water vapor starts to condense), thus defining saturation humidity. Essentially, the partial pressure of the water vapor is equal to the vapor pressure of pure water at equilibrium conditions of the same pressure and temperature. Air relative humidity (*RH*) is defined as:

$$RH = 100 \, P_{\text{w}} \, / \, P_{\text{sat}} \,, \tag{5}$$

where P_{sat} represents the saturation pressure of water vapor at a given temperature.

Families of relative humidity curves and wet bulb temperature lines are plotted on coordinates of absolute humidity versus dry bulb temperature to form a basic psychrometric chart. Such a chart, or similar ones constructed for specific system pressures, is a useful tool for determining the properties of air. Other properties, such as humid heat, humid volume, and latent heat are usually plotted on the chart and are used for calculating mass and energy balances in drying calculations. Geankoplis (1983) and other texts illustrate the fundamentals and use of these charts.

The *dew point* is the temperature at which a sample of air becomes saturated at the pressure and absolute humidity of the sample. The dew point thus lies on the saturation curve.

The *enthalpy* (*H*), in kilojoules per kilogram of dry air, of an air-water vapor mixture is the sum of the sensible heat and the latent heat and is given as:

$$H = (1.005 + 1.88W)(T - T_0) + 2{,}501.4W \,, \tag{6}$$

where T_0 is the reference temperature.

The *adiabatic saturation point* is the temperature a sample of air reaches in steady state with a large amount of water. The *wet bulb temperature* is the steady-state, nonequilibrium temperature reached under adiabatic conditions for a given sample of air in contact with smaller amounts of water (Geankoplis, 1983). From thermodynamic considerations, the two temperatures are essentially the same for air-water vapor mixtures. Accordingly, the wet bulb temperature is used to determine air properties on the psychrometric chart.

The important process parameters are the air dry bulb temperature, air relative humidity, retention time needed to accomplish the required drying, airflow pattern and velocity, product temperature, and desired product moisture content before and after drying. Given these basic parameters, mass and energy balances and process control can be established. These and many other parameters are determined before operation by means of batch drying tests.

IDEALIZED BASIC DRYING PROCESS

An idealized basic drying process for a cereal pellet dryer is illustrated in Figure 2 on a schematic representation of a psychrometric chart. Supply air (the air before it goes through the product bed) is at conditions of point B. As the air goes through the product bed, it releases the heat necessary to evaporate the water in the product and carries away the vaporized water. The air basically follows the adiabatic saturation line. The actual properties of the air and of the drying process may deviate significantly from the idealized process if conductive or radiative heat transfer interferes. For pellet drying, the assumption of adiabatic conditions is accurate enough for practical purposes in describing the drying process.

At point C, the air has completed its pickup of moisture from the product. The cooling effect, known as *evaporative cooling*, significantly affects the drying process and heat requirements. For cereal pellet dryers, a portion of the circulated air is exhausted at this point. Fresh ambient air at point A is added to the rest of the air and mixed to form air at point D. The specific consequences of this mixing in terms of energy consumption and product quality are discussed later. This mass is now reheated along line DB to become supply air.

The *dry bulb temperature* should be hot enough to provide a reasonable drying rate and yet is limited by possible deleterious physical and

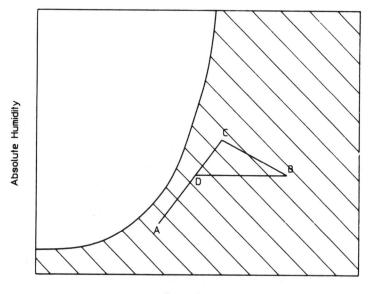

Temperature

Figure 2. Idealized drying process on a schematic representation of a psychrometric chart. Conditions of air at various stages: A, fresh ambient air; B, supply air; C, air after moisture pickup; D, mixed air.

chemical reactions. The air relative humidity should be high enough to avoid serious case hardening and yet low enough to provide as much driving force as possible. The purpose in making these choices is to maximize heat and mass transfer before exhausting. Thus the choice is based on minimizing energy use without compromising product quality. Also, the relative humidity must be set low enough to prevent condensation on the surfaces of either the product or the dryer, which could create problems for drying and sanitation.

The choice of *relative humidity* also depends on the required final moisture content of the product. The water activity relationships for wet products have been well discussed by Karel (1975). The basic concept is that of *equilibrium relative humidity*, the specific amount of water associated with the product at a given temperature and air relative humidity. If no changes are made in air conditions, equilibrium is reached and there is no net evaporation of moisture from the product. Figure 3 is a schematic family of curves drawn to show the relationship between air relative humidity and product moisture content at given temperatures.

The *retention time* for a given product should be determined by testing, although it can be estimated by various equations (Geankoplis, 1983). The time varies with air temperature and relative humidity, as already discussed. It is important to optimize retention time because the retention determines the dryer holdup (i.e., the amount of material

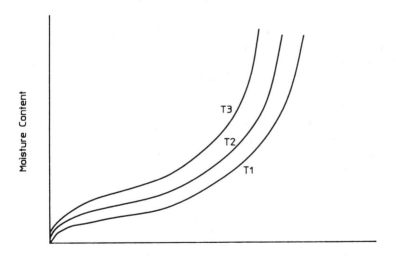

Relative Humidity or Water Activity

Figure 3. Schematic family of food moisture isotherms. T = temperature.

in the dryer at any given time) and dryer bed depths, which are crucial to dryer operation and also have an impact on energy consumption.

Airflow pattern and *velocity* must be chosen to provide adequate flow around each particle and to accomplish appropriate drying rates. Limits are set by particle incipient fluidization and bed properties.

The *product temperature* going into the dryer is needed to determine the mass and energy balances. The desired *product moisture* in and out determines what type of drying mechanism is important and also sets the limits for air relative humidity.

PRODUCT PARAMETERS

The physical and transport properties of the material to be dried are briefly discussed in this section in order of importance. Certain properties are more significant for drying with some products or formulations than with others.

Product feed rate is a major factor determining dryer size. *Product moisture data* in and out, which are determined, as mentioned above, by the preceding and following processing steps, are needed to determine evaporative loss. A desired *moisture loss pattern*, as established by a drying curve from preliminary batch testing, is important in determining dryer conveyor speeds and machine sizing.

Product bulk density must be known in order to determine dryer holdup and guide heights. The bulk density does change due to shrinkage, but this can be offset by the loss of water, and for many pellets bulk density remains approximately constant.

Tests of *pressure drop* versus bed depth provide information on airflow and energy requirements. Physical characteristics such as *particle size and shape*, stickiness, and internal viscosity affect the drying rate and further processing. The equilibrium relative humidity curves for the product at operating temperatures determine drying potential. Critical moisture content data for a specific pellet formulation and size help in airflow considerations and sizing determinations. For most cereal pellets, the critical moisture content is around 32%, wet basis. For coated products, the amount of coating applied, the moisture content, and the characteristics of the syrup as it dries must be known for proper dryer sizing.

Selecting a Dryer

General considerations that apply in every case of dryer selection are that the equipment manufacturer should understand the larger process of which the drying in question is just a part and should be able to provide good service; the dryer should be properly designed for good control of the temperature and humidity of the air, which are essential for efficient operation; and the quality of workmanship (especially for food industry equipment) should ensure safe, wholesome, high-quality products. Sanitary design, rounded-corner welding, and avoidance of interfacing equipment surfaces are some of the details that play important roles in building the best equipment.

DRYER DESIGN

The basic principle of dryer design is to solve the coupled heat and mass transfer equations (Lapple and Clark, 1955; Zanker, 1976) to determine dryer area and airflow requirements. These are translated into dryer sizing requirements, given the product and process parameters previously discussed. Each manufacturer has proprietary design schemes and follows specific established practices.

It may be useful to mention a few physical design details. The conveyors should allow good airflow and yet provide good product support. Special conveyor-surface coatings may be used in case a material is sticky. Airflow pattern is crucial: all product must "experience" uniform airflow and conditions. The selection of a fan or blower is important to ensure adequate volume, static pressure, and distribution of air. Coad and Sutherlin (1974) and Van Arsdel and Copley (1964) discuss basic criteria in fan selection. The heating source most commonly used is steam with finned coils. Direct gas-fired equipment is also used but is limited in the maximum relative humidity attainable. Hot water coils have also been used, but decreased capacity (due to fouling) and large coil sizes detract from their usefulness.

THE SELECTION PROCESS

Lee (1974, 1975) described five steps in dryer selection. First, state the problem as accurately and completely as possible. Second, collect all available data pertinent to the problem. Third, determine the critical factors. Fourth, narrow the choices to possibly three or four drying systems based on the most important variables. Fifth, evaluate all the critical information, including final product and cost considerations, to make the final decision.

Used as a guideline, this scheme works well. For example, the purposes for drying cooked cereal pellets include preparing the material for subsequent processes and removing water needed during previous operations, both of which Lee (1974) lists.

The necessary design criteria have been mentioned in the previous sections on the principles of drying, and the critical processing factors have been outlined. However, marketing, operations, and economic considerations have not been covered. Sapakie et al (1979) carefully examined the operations and economics of drying in the food industry. Their results indicate that capacity is a key factor and that forced-air dryers are a good choice given operational and fixed-cost comparisons. Further details on economic analysis may be found in the literature (e.g., Lapple and Clark, 1955; Peters and Timmerhaus, 1968; Perry and Chilton, 1973; Maris, 1979).

TYPES OF DRYERS

In cereal processing, some *rotary dryers* are used to predry pellets. Rotary dryers were common as pellet dryers, but the disadvantages of inadequate residence time control and nonuniform product moisture and drying in many cases dictate a different selection.

Some processors use *turbo dryers*. These are well suited for gentle treatment of products. Air flows across the product bed. Fines generation is minimal. Turbo dryers can be used to dry cereal pellets before flaking or puffing; however, they are best used with thin beds. Lapple and Clark (1955) and Lapple et al (1955) provide a good summary of turbo dryer parameters and operation.

Single-pass conveyor dryers are also used, particularly for those cereals or cereal products that require baking or toasting as primary process steps, and where moisture removal is seen as a concurrent or secondary issue. However, multipass conveyor dryers are probably the most widely used for the main drying of cereal pellets. Accordingly, the discussion that follows centers on these dryers.

Multipass conveyor dryers (Figure 4) have some distinct advantages over other designs. Products may be dried in thinner beds at the times that drying rates are the highest, with heavier belt loadings at lower drying rates. The first pass can be a single layer or one of very shallow product depth for products that are sticky, providing quick surface drying to minimize lumping. The reorientation of product from pass to pass changes the orientation of individual product pieces and thus their exposure to the drying air, resulting in more uniform moisture distribution. With variable speed drives on each conveyor pass, the drying curve for a particular product can be optimized or modified as desired.

When the most product moisture is removed at the thinnest bed depths, the drying ability of the circulating air does not change much as the air travels through the product bed, undergoing relatively small temperature and moisture changes, which results in more uniform treatment of all product in the bed. In addition, multipass dryers generally require less floor space than other types because the required aerated belt can be divided into several stacked conveyors for a smaller area (Figure 4).

Figure 4. Typical multipass dryer (views from opposite ends). (Courtesy Food Engineering Corporation, Minneapolis, MN)

PREDRYING

Pellets may be predried to improve material handling. Sometimes this is accomplished with a simple fan over a conveyor belt; other times, rotary dryers are used. Temperatures range from ambient air to 333°K (140°F). The object is to cool the product, lower water diffusivity to inhibit sweating, and take off some surface moisture. Actual moisture loss is typically 2-5% (wet basis), with product coming in at 30-35% moisture. Maris (1979) and Van Arsdel and Copley (1964) give more specific information on rotary dryer operation, design, and use.

PRODUCT MOISTURE

The desired result of drying is to prepare the cereal pellet for further processing. Moisture content is one of the most important variables. Many methods and machines are available for measuring product moisture, and the processor should choose among them based on time, cost, accuracy, reproducibility, and reliability.

Cooked cereal pellets are dried chiefly to ensure proper flaking, puffing, forming, and toasting. The entering and exiting product moistures are determined by the moisture content necessary for these operations. The product and process parameter values vary depending on the ingredients, end product, and dryer design. Product moistures, especially, depend on pellet stickiness. Wheat and bran formulations tend to be less sticky than those of corn and rice. Cooked pellets that will be "bumped" using flaking or other rolls typically enter the dryers at 30-33% moisture (wet basis) and exit at 16-22%. Pellets that are puffed in guns or other machines typically enter the dryers at 30-32% moisture (wet basis) and exit at 10-12%.

Drying also influences physical characteristics of the pellets such as uniformity of moisture content, surface dryness, and dough viscosity. Even with full temperature and relative humidity control, a moisture gradient persists from a high point at the center of the pellet to a low point at the surface. The average moisture content is what is usually measured if the sample is properly prepared. Tempering is used to reduce the moisture gradient. Additional technical information regarding cereal processing may be found in such resources as Daniels (1974).

AIRFLOW AND DRYER CLIMATE

Pellet dryers are usually designed to provide airflow velocity within the range of 0.356-1.524 m/sec (70-300 ft/min). Fan static pressures range from 124.3 to 994.5 Pa (0.5- to 4-in. water column), depending on the airflow pattern and fan used. Dry bulb temperatures for the supply air (air before passing the product) range around 335–383°K (180–230°F) and may be as high as 416°K (290°F). Supply air relative humidities in the range of 8–15% may be used for fairly efficient drying. The air humidity should be chosen to optimize performance. Figure 5 shows energy need as a function of relative humidity.

BED COVERAGE AND DEPTH

Uniform bed coverage and depth are crucial for proper and efficient drying. Product bed depth (for a given dryer size) is most affected by product feed rate, bulk density, and retention time. These parameters should be determined as accurately as possible to design for appropriate bed depths. Pellet dryer capacities are usually rated at the inlet moisture and generally range from 454 to 5,448 kg/hr (1,000–12,000 lb/hr). Bulk densities may range from 320 to 800 kg/m³ (20–50 lb/ft³). Retention time ranges from 10 to 90 min.

Infeed and product distribution devices are an important part of dryer design. Chutes, belts, and vibrating pans convey the product to the dryer, where either the discharge is specifically designed to distribute the product on the belt or a plow, rake, or sweep is used to level the

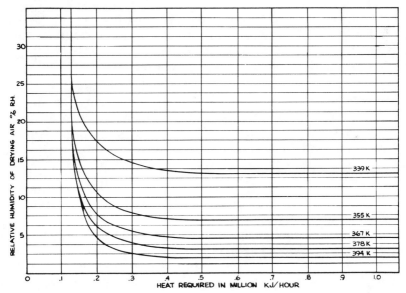

Figure 5. Dryer energy requirements as a function of climate relative humidity.

product bed. Rotary pickers inside the dryer can be used to rearrange the product bed for a different airflow orientation as well as to help minimize clumping and ensure uniform bed depth and distribution. At the dryer exit, specially designed devices, including chutes, augers, and vibrating pans, can be used to discharge the product.

COATED PRODUCTS

Dryers used for sugar coating or other topical coatings are usually one- to three-pass conveyor dryers. The airflows are typically milder than those for pellet dryers, but the temperature may range both lower and higher. The moisture reduction is in the range of 8–11%, wet basis. Special care in design must be taken to ensure proper hardening and solidifying of the coating given processing parameters. It is also important to prevent mechanical action or disturbance during critical parts of drying (Johnson and Peterson, 1974).

DRYER CONTROLS

Dampers

Uniform airflow and air conditions are essential to efficient drying. Airflow is normally controlled by proper sizing and operation of fans and dampers. Dampers are used to create a slightly negative pressure inside the dryer as well as for climate control. They are usually placed at the inlet and exhaust air openings. The negative pressure provides protection from high working temperatures and ensures that the supply air passes through the beds. A pair of fans can also be used to accomplish this, but normally an exhaust fan and dampers suffice.

Dampers are also used elsewhere in the dryer. Some may change the amount of air recirculated. Inverters for changing fan speed or selecting specific motor banks can also be used for airflow modification. Other dampers ensure upflow or downflow through one or more passes. Dampers and flaps are also used to separate aerated from nonaerated sections and to prevent air leakage from a cooling section into the hot, humid climate of the dryer or vice versa.

Temperature

A constant dry bulb temperature for the airflow improves the uniformity of drying. The temperature is usually sensed by a thermocouple or resistance temperature device. A signal is sent to a controller that in turn actuates a steam, hot water, or gas valve. Temperature may be controlled automatically with a proportional integral derivative (PID) loop controller or various computer controllers. The control system can be pneumatic, current-operated, or a combination. The dry bulb temperature should be as high as practical because of the savings in energy consumption due to the ability of the air to hold more moisture at higher temperatures (Zagorzycki, 1979).

Humidity

Temperature control alone was used for a long time. Higher energy costs and the desire to improve dryer performance led to the development of full climate-control systems. Such systems measure the air humidity and temperature and feed the measurements into control loops to actuate dampers and valves that control heat input and exhaust and inlet airflows.

Such sophisticated control systems improve product quality and decrease operating costs for the cereal processor (Zagorzycki, 1983). Product moisture and quality can be kept uniform from morning to evening and from summer to winter, regardless of ambient climate changes in the plant or outside weather changes. Substantial changes in product feed rate do not significantly affect the uniformity or final moisture content of the product. Less dryer exhaust air and plant makeup air are required, and significant energy savings may be realized.

The sensors can be located in either the supply or the leaving air relative to the product bed. Although either placement provides good dryer control and economic savings (Zagorzycki, 1983), locating the sensors in the leaving (exhaust) airstream is particularly advantageous. First, because the humidity and volume of the exhausted air are the major determinants of energy use, placing the control at the exhaust point ensures optimum performance. Second, in cereal pellet drying, the most important variable in dryer operation is feed rate. Zagorzycki (1983) clearly shows how humidity control is economical for operation at partial capacity. Locating the sensor on the exhaust air side saves a significant amount of energy by adjusting the climate in response to changes in the feed rate. Figure 6 is a schematic representation of such an application and also illustrates climate control response to changes in dryer capacity.

Operation of the control system is straightforward. When product enters the dryer, the dampers are closed and all the air is recirculated. Humidity builds up as moisture evaporates, and soon the exhaust dampers begin to modulate open. The inlet dampers are usually slaved to allow enough air to flow in to replace that exhausted. At steady state, both the dampers and steam valves rest at some position. Proper tuning of the control loops, whether controlled by PID loop controllers

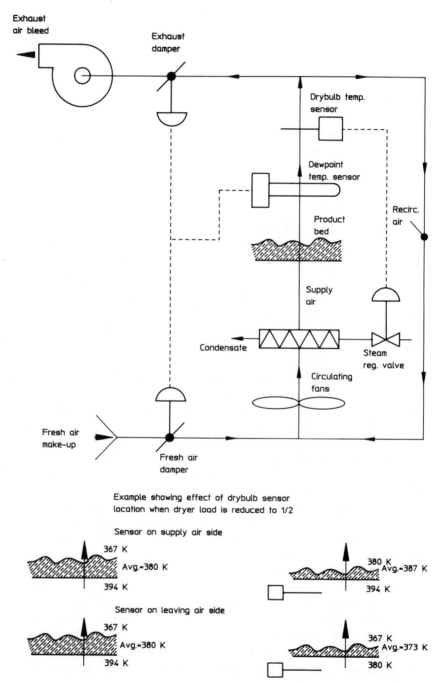

Figure 6. Schematic representation of a climate control system for cereal dryers.

or other means, provides close control even during start-up and shutdown.

Humidity control can be modified to include steam injection as a source of moisture. In terms of water vapor, this is not usually necessary because moisture is available from evaporation. However, steam injection is useful for dryer preconditioning and may be helpful for sensitive products, particularly if the dry bulb and humidity loops are not coupled.

CLEANING AND SANITARY DESIGN

Dryers can be designed for various cleaning methods. Routine cleaning is usually done dry with brushes and heat. Special design features can be incorporated for wet washing, such as waterproofed or isolated electrical and drive equipment, sloping floors, and drains. Clean-in-place systems are also used. Care should be taken to match construction materials and cleaning chemical properties.

Original dryer design should conform as closely as possible to such standards as ANSI-ASME F2.1 and F2.la (ASME, 1975, 1976) and Baking Industry Standard No. 14 (Anonymous, 1968). The basic principle behind sanitary design is accessibility of equipment parts to sight and touch. This translates into easy and clear access to all parts of the dryer, appropriate (i.e., stainless steel) food product surfaces, rounded-corner welding, and no nonessential horizontal surfaces. Painted surfaces should be avoided to prevent chipping and rust contamination. The dryer design should minimize product damage and spillover. The heating system should be designed for accessibility and ease of cleaning as well as energy efficiency. Belt brushes and fines collection systems may be included if used carefully and checked frequently.

SAFETY FEATURES

Cereal dryers should follow industry standards such as those mentioned in the previous paragraph to safeguard both the operators and the equipment. Fire sprinklers or sprayers can be designed to fit each dryer. Fan guards and enclosures for moving parts are important. Proper arrangement and guarding of the heating system should not be overlooked. Torque or motion sensors on the conveyors will prevent machine damage and protect operators. Guards and emergency stops are important for infeed and discharge equipment. The door latch should be designed so that an operator accidentally caught inside could get out.

References

Anonymous. 1968. Baking Industry Standard No. 14. In: Sanitation Standards for the Design and Construction of Bakery Equipment and Machinery. Baking Industry Sanitation Standards Committee, Chicago, IL.

ASHRAE. 1977. ASHRAE brochure on psychrometry. American Society of Heating, Refrigerating and Airconditioning Engineers (ASHRAE), New York.

ASME. 1975. ANSI-ASME F2.1: Food, drug, and beverage equipment. The American Society of Mechanical Engineers (ASME), New York.

ASME. 1976. ANSI-ASME F2.1a: Addenda to food, drug, and beverage equipment ANSI-ASME F2.1 (1975). The American Society of Mechanical Engineers (ASME), New York.

Coad, W. J., and Sutherlin, P. D. 1974. A new look at fan system curves. Heat. Piping Air Cond. 46(11):47-51.

Crank, J. 1956. The Mathematics of Diffusion. Clarendon Press, Oxford, England.

Crank, J. 1958. Some mathematical diffusion studies relevant to dehydration. Pages 37-41 in: Fundamental Aspects of the Dehydration of Foodstuffs. Society of Chemical Industry, London. (Also in Van Arsdel et al, 1973)

Daniels, R. 1974. Breakfast cereal technology. Food Technol. Rev. 11. Noyes Data Corporation, Park Ridge, NJ.

Geankoplis, C. J. 1983. Transport Processes and Unit Operations, 2nd ed. Allyn and Bacon, Inc., Newton, MA.

Johnson, A. H., and Peterson, M. S. 1974. Encyclopedia of Food Technology. AVI Publishing Co., Westport, CT.

Karel, M. 1975. Dehydration of foods. Page 323 in: Principles of Food Science, Part II: Physical Principles of Food Preservation. O. R. Fennema, ed. Marcel Dekker, New York.

Lapple, W. C., and Clark, W. E. 1955. Drying: Methods & equipment. Chem. Eng. (N.Y.) 62(10):191.

Lapple, W. C., Clark, W. E., and Dybdal, E. C. 1955. Drying: Design & costs. Chem. Eng. (N.Y.) 62(11):177.

Lee, D. A. 1974. Five steps to simplify selecting a dryer. Part I. Chiltons Food Eng. 46(11):81-83.

Lee, D. A. 1975. Five steps to simplify selecting a dryer. Part II. Chiltons Food Eng. 47(3):63.

Maris, J. 1979. Selecting the right dryer is critical to cost control. Chem. Process. (Chicago) 42(12):36.

Perry, R. H., and Chilton, C. H. 1973. Chemical Engineers' Handbook. McGraw-Hill Book Co., New York.

Peters, M. S., and Timmerhaus, K. D. 1968. Plant Design and Economics for Chemical Engineering, 2nd ed. McGraw-Hill Book Co., New York.

Sapakie, S. F., Mihalik, D. R., and Hallstrom, C. H. 1979. Drying techniques: Drying in the food industry. Chem. Eng. Prog. 75(4):44-49.

Sherwood, T. K. 1929. The drying of solids. Ind. Eng. Chem. 21(1):12-16. (Also in Van Arsdel et al, 1973)

Slesser, C. G. M., and Cleland, D. 1962. Surface evaporation by forced convection. I. Simultaneous heat and mass transfer. Int. J. Heat Mass Transfer 5:735-749.

Van Arsdel, W. B., and Copley, M. J., eds. 1964. Food Dehydration, Vol. II: Products and Technology. AVI Publishing Co., Westport, CT.

Van Arsdel, W. B., Copley, M. J., and Morgan, A. I., eds. 1973. Food Dehydration, Vol. I: Drying Methods and Phenomena. AVI Publishing Co., Westport, CT.

Wilgen, F. J. 1979. Dryer model for cooked cereal pellets. Technical paper, National Conference and Exhibit, Chicago. Instrument Society of America, Pittsburgh.

Williams-Gardner, A. 1971. Industrial Drying. CRC Press, Cleveland, OH.

Zagorzycki, P. E. 1979. Drying techniques: Automatic control of conveyor dryers. Chem. Eng. Prog. 75(4):50-56.

Zagorzycki, P. E. 1983. Industry drying practices: Automatic humidity control of dryers. Chem. Eng. Prog. 79(4):66-70.

Zanker, A. 1976. Two nomographs calculate all basic atmospheric dryer data. Chiltons Food Eng. 48(8):58-59.

Chapter 5

Unit Operations and Equipment
III. Tempering, Flaking,
and Toasting[1]

ELWOOD F. CALDWELL   FRED J. SHOULDICE
ROBERT B. FAST   DONALD D. TAYLOR
SUSAN J. GETGOOD   WILLIAM J. THOMSON
GEORGE H. LAUHOFF

Tempering

The term *tempering* in a breakfast cereal process denotes a unit operation different from the operation with the same name in flour milling. In flour milling, tempering refers to the addition of a small amount of moisture to the cleaned grain before milling. The added water toughens the bran coat so it will remain in larger pieces during the grinding and sifting of the flour fractions. In a breakfast cereal process, tempering usually follows a drying or cooling step and is the period during which the cooked grain mass or cereal pellets are held in collection bins to allow the equilibration of moisture within and among the particles and the development of desired flakability or shredability.

The effect of tempering on final product quality depends to some degree on drying conditions. Proper drying conditions greatly cut down on temper time. Manufacturers should resist the temptation to raise dryer temperatures to push up production rates. Prolonged tempering

[1]Principal contributors: of the section on tempering, Robert B. Fast; on flaking, Robert B. Fast, George H. Lauhoff, Donald D. Taylor, and Susan J. Getgood; on toasting, Robert B. Fast with Fred J. Shouldice (rotary oven), with William J. Thomson (band oven), and with Donald D. Taylor and Susan J. Getgood (air impingement).

cannot make up for poor dryer operation; improper drying only compounds itself during tempering.

The drying step in a cereal process usually leaves the cooked grain or pellets with a nonuniform distribution of moisture. The centers of the particles may contain more moisture than the surfaces, as in grits dried from about 30% moisture down to 17% for flaking. In the case of shredded whole wheat, on the other hand, where the grain has been cooked in excess water rather than a limited amount that is all absorbed during cooking, the step following cooking is primarily a cooling one, with some evaporation of surface moisture. This results in a moisture disparity opposite to that in grits, with the interior being drier than the areas closer to the surface. In both cases, tempering time provides for equilibration of moisture within the particles. A third situation concerns grain that is cooked to about 30% moisture in limited water (which is all absorbed) and then cooled rather than dried as such. It remains in the 28–30% moisture range, but other changes take place during tempering time (such as the starch retrogradation explained later) that result in improved shredding characteristics.

In all three of these situations, cooling the grain before tempering greatly improves the quality of the tempered material. If grain dried to 17% moisture for flaking is binned at temperatures much over 100°F, its color darkens during tempering. The additional color stays with the product through flaking and toasting and shows up in the finished flakes. The same is true for grain cooked and cooled to 30% moisture for shredding. Wheat cooked for shredded wheat and binned at elevated temperatures remains overly sticky and will not shred properly or produce the desired long, straight shreds that result in attractive biscuits. Thus, the importance of cooling before tempering cannot be overemphasized.

Tempering bins come in a variety of sizes and shapes, depending on the product. Some bins for flaking grits are merely large, open-top, rectangular containers (Figure 1). The bottom of such a bin is a slow-moving conveyor belt. The vertical sides and one end are mounted at the outside edges of the conveyor belt, and the other end of the bin typically consists of a set of rakes inclined back toward the center of the bin about 15°. The rakes are driven at moderate speed so that those closest to the pile of grits in the bin are moving upward. Such bins are typically about 10 ft wide, 20 ft long, and 5 ft high. The top may be closed but need not be.

These bins are loaded from a central point from an overhead conveyor running the length of the bin. The first product loaded into the bin is also the first to be discharged at the end of tempering. The forward movement of the conveyor bottom toward the rakes at the end of the

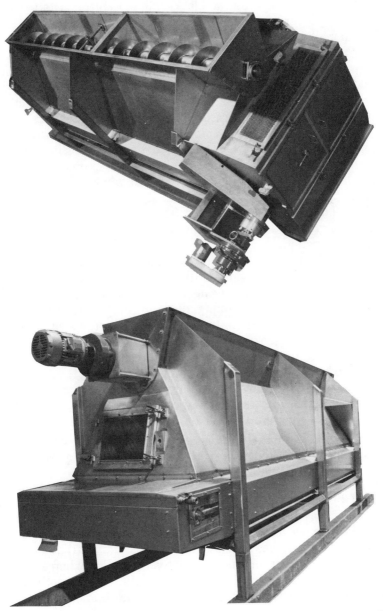

Figure 1. Tempering bin for cooked cereal grits or pellets. The top view shows the inlet and discharge conveyor; the bottom photo is an end view. (Courtesy Food Engineering Corp., Minneapolis, MN)

bin discharges the product. Grits have a tendency to block in a solid mass when allowed to stand in a bin like this. The rakes moving across the face of the pile break up the blocks and loosen the grits from the bin surfaces. The grits drop onto a conveyor to move to the next step, which is usually sifting to ensure that only grits of the correct size go on for further processing. Sifting removes any fines generated and separates the "overs," or large pieces, for more breaking and resifting.

In another, much simpler type of tempering setup, the tempering facility is merely an additional section of the dryer. Because humidity is controlled as well as temperature, no case hardening occurs during drying, which greatly reduces temper time to equilibrate the moisture within particles. In this type of system, a cooling section is incorporated into the dryer between the drying and tempering sections.

In a third type of tempering setup, used in whole-grain shredding, the cooked and cooled grain is held in large, cylindrical bins before shredding. The bins are usually fed from above by a conveyor, and the tempered grain is discharged from the cone-shaped bottom through a central opening.

Tempering is not as simple and straightforward as might appear from the foregoing description. Until recently, little was understood about the mechanisms of tempering. In the early days of the industry, tempering was regarded as simply allowing time for the moisture in the particles or grain kernels to equilibrate. Dryers dried the grain with hot air and lacked modern humidity controls. Particle centers were always wetter than the surfaces, a situation that was not conducive to further processing. In flaking, for instance, the dry surface prevented grits from being drawn into the nip of the flaking rolls. If they were drawn into the rolls by chance, the moist centers were often so gummy that the flakes balled up on the flaking roll knives. In fact, one of the early inventors in the industry, Dr. John Harvey Kellogg, was not able to make wheat flakes until his staff inadvertently allowed cooked wheat to stand overnight before flaking.

When starch retrogradation theories began to appear to explain the staling of bread, breakfast cereal scientists wondered whether this phenomenon was also at work in the changes observed in grain or grits during tempering (White, 1979). Cooled, untempered wheat can be shredded but produces short, discontinuous strands that are tough and curly and do not produce attractive biscuits, whereas wheat tempered for 8–28 hr at 15–30°C makes straight, long, smooth shreds. White (1979) demonstrated that these desirable characteristics could be induced in the cooked wheat in less time by lowering the temperature of the grain during the temper period. This is consistent with the fact that starch retrogradation (and the associated staling of bread) is maximized at refrigerator temperatures. Jankowski and Rha (1986) further reported on the retrogradation phenomenon, pointing out that the firmness of grain due to starch crystallization (as indicated by the elastic modulus) increases with time and is greater at lower storage temperatures. Lowered storage temperature not only accelerates starch crystallization and thus the firming of cooked grain but also seems to generate a different organization of starch molecules.

Tempering is thus a significant step in any cereal process. Because of its effects on final product quality, it should not be considered as merely a holding period that can be adjusted to accommodate shift rotations, as has frequently been the case in the industry in the past.

Flaking and Flaking Mills

Flaking mills consist of two rolls, one of which is adjustable so that the distance between them, or roll gap, can be set to produce a flake of the desired thickness. Either one or two drive motors may be used. A scraper knife on each roll removes the flakes, which are then conveyed to a toasting oven.

When flaked cereals were first developed and marketed, small (e.g., 9 in. in diameter), chilled-iron, smooth-ground flour mill rolls were often used for flaking. It soon became evident that such rolls were inadequate for flaking cereals. Larger rolls proved to be better, and a diameter of 20 in. became generally accepted as the minimum for ready-to-eat cereal flakes. However, it was relatively more difficult to control the surface temperature of the rolls of this diameter that were of the traditional solid construction. A water-cooled flaking roll patented by Lauhoff (1913) was heralded as the ideal solution to controlling temperature to maximize production and quality of flaked cereals.

TYPES OF FLAKING MILLS

Today there is no one "standard" mill used by cereal manufacturers and no real theoretical support for any particular design or specification of roll face width, diameter, hardness, or speed. The most popular sizes (diameter by width, in inches) are 20 × 30 and 26 × 40. Other standard sizes are 32 × 40, 24 × 40, and 20 × 24.5 for ready-to-eat cereals and 14 × 28 for rolled oats and wheat.

Several manufacturers in the United States and abroad make flaking mills (Figures 2-4). Selecting among the available models involves comparing features such as type of roll material (including rated surface hardness and depth), type of cooling system, size of roll journals (shafts), and size and type of bearings used on the roll shafts.

ROLL CONSTRUCTION

Flaking rolls used in the cereal industry are usually made from chilled iron or centrifugal alloy-iron castings or from a solid steel forging. The roll surfaces must be hard to resist the abrasive action during flaking. Centrifugal cast rolls tend to be more uniform in hardness throughout the depth of the hardness zone than chilled iron rolls.

COOLING

Chilled iron rolls are hollow (chamber-bored) so that they can be cooled by passing water through the interior of the roll. Rolls made from solid steel forging are grooved on the surface with multispiral grooves. Coolant flows through these grooves at 10–100 gal/min. A replaceable outer shell or rim covers the grooved casting. This rim may be hardened through the full 1-22/32 in. thickness or flame-hardened to a depth of 5/16 in. Its hardness is usually Rockwell C 55-60.11.

Figure 2. Cereal flaking roll stand with 26 × 40 rolls. (Courtesy Wolverine Corp., Merrimac, MA)

Rotary unions at each end of each roll allow coolant to flow into and out of the rolls. Water chillers are commonly used to control cooling-water temperature. Without such chillers, roll performance can vary from winter to summer in northern climates because of the normal seasonal changes in cold water temperatures.

Other rolls have a series of drilled passages adjacent to the roll surface. Cooling water is circulated through these passages to control roll-surface temperatures.

In actual use, roll surfaces start at room temperature and gradually warm up because of the friction associated with the flaking process. It may take a half hour or more for a roll stand to reach normal operating temperature, at which time cooling is turned on to regulate the surface temperatures. Surface temperatures in the range of 110–115°F (43–46°C) are normal and about optimum for rice and corn products.

As the rolls heat up, they expand, which naturally reduces the gap between the rolls and thereby changes the thickness of the flakes produced. Changes in the amount of product fed to the rolls also affect the surface temperature and thus the thickness of the flakes.

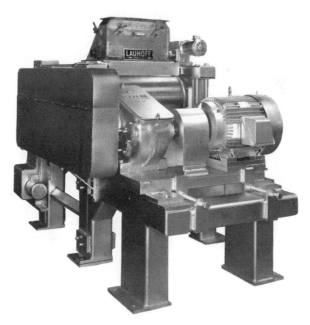

Figure 3. Cereal flaking roll stand with 20 × 30 rolls (rear view). (Courtesy Lauhoff Corp., Detroit, MI)

ROLL STANDS

The stands on which flaking rolls are mounted vary from manufacturer to manufacturer. Some are made of welded tubular steel on which the bearings for the rolls are mounted. Others are made from solid, hot-rolled steel plates. The stands must be strong enough to withstand the pounding they take from the flaking process and rigid enough to keep the rolls in tram (parallel). Ease of cleaning is another important consideration in stand design; access to all surfaces is needed, and hidden channels or other surfaces where food could collect should be avoided.

One roll in the stand is mounted in a fixed position, and the other roll can be moved. The gap-positioning device regulates the spacing between the rolls by either pushing the movable roll toward the fixed roll or pulling it away. Sometimes the movable roll is spring-loaded for overload protection to prevent damage to the roll surfaces.

SETTING THE GAP

Gap-positioning devices generally fall into three categories: hand-operated screws, motor-driven screws, or hydraulic cylinders. All three types are in operation in cereal plants today.

Figure 4. Three stands of 20 × 30 cereal flaking rolls with 60-hp drives, roll feed, and electromechanical roll adjustment. (Courtesy Lauhoff Corp., Detroit, MI)

The roll gap is normally set according to visual examination of the flakes being produced. The gap is usually set after the product feed starts, to minimize the chance that the rolls will touch while they are running. Newer technologies, including electronic or sonic gap positioning, offer absolute gap control and lessen or eliminate the role of the operator in controlling product thickness. Laser roll positioners are also now available.

In hand-operated models, handwheels on each end of the movable roll are turned to tighten large screws that push the bearing blocks on each end of the roll toward the fixed roll, thus closing the roll gap. A lever 4–6 ft long may be needed to apply enough torque to the handwheels to position the gap for proper flaking while product flows to the roll nip. Each end of the movable roll must be positioned in this manner individually.

The use of power torque setters is an improvement over hand setting. Here the handwheels are replaced by a slow-speed motor drive. By simultaneously pressing both "in" or both "out" buttons, one can move both ends of the movable roll at once.

Systems employing hydraulic cylinders are also used to push or pull the ends of the movable roll toward or away from the fixed roll. Hydraulic pressures of 2,000–3,000 psi are common. These pressures are developed by a separately driven hydraulic fluid pump system.

KNIVES

Steel knives, usually about 3/8 in. thick with a 0.5- to 0.75-in. bevel to a sharpened edge, are used to shave off flakes that happen to stick to the roll surface. The bevel side of the knife faces downward. Knives are typically made of tool steel, with a hardened tip or inlaid edge. It is important for the knife edge against the roll to be less hard than the roll surface, or the knife will wear down the roll surface. Knives should be set against the roll surface with only enough pressure to peel flakes from the roll. Excessive pressure will heat the knife edge and roll surface and result in rapid wear and dulling of the knife edge.

Knife blades are mounted in holders, which can be moved toward or away from the roll by means of screws with handwheels or by pressure applied by very low-pressure air cylinders. The holders are constructed in such a way that the blade rests on a fulcrum point on the bottom half of the holder. Screws spaced on about 3-in. centers come down through the top half of the holder and rest on the top of the blade itself. By selective tightening of these screws, the blade can be flexed to a tighter position against the roll to clean a particular area of the roll surface better. Knife holders are also available that oscillate back and forth against the roll to provide better cleaning of the roll surface.

DRIVES AND DIFFERENTIALS

Drives for flaking rolls range from one motor and reducer for each roll to a single motor and reducer plus a drive chain or belts for both rolls. Power requirements range from 40 to 100 hp, depending on the product. Wheat and oats generally require less power than corn or rice flakes. The capacity of a roll stand for corn flakes can range from 400 to 3,000 lb/hr, depending on the size of the rolls and the choice of drive and horsepower.

Roll speed differentials are used by some but not all cereal manufacturers. The thinking behind having one roll running faster than the other is that the faster roll has an "ironing" effect on the flakes, stretching or elongating them. The two rolls working together at the same speed simply flatten the material and produce more uniform flakes. Roll speeds are normally in the range of 100–300 rpm, and speed differentials of 4–8% or higher are not uncommon.

FEEDERS

Three alternate types of feeders are used to feed grits or pellets to the flaking rolls. The first and most common is a feed hopper with a bottom that consists of a corrugated roll instead of coming to a V shape. The lower parts of the V sides in the feed hopper move up or down to adjust the flow of grain or pellets to the rolls. The corrugations are deep enough to accommodate the pellets without damaging them. The second type is a sloping, vibrating pan that meters a uniform volumetric flow to the roll nip, and the third is a slow-moving belt that meters a volumetric or gravimetric flow to the roll nip.

ROLL OPERATION

Products in the 15–17% moisture range generally flake with little or no difficulty. Products lower in moisture, if not surface-conditioned, are often not grabbed by the roll nip and therefore do not feed through the rolls. Products with more than 17% moisture often are too soft and sticky and will not release from the roll surfaces. These high-moisture flakes also often stick to one another, causing clumps in the finished product.

Operating a flaking mill is more difficult than appears at first sight. Setting the proper roll gap, maintaining the proper roll surface temperature, and achieving the optimum setting of the roll knives are all tasks that require training and close attention on the part of the operator.

Toasting and Toasting Ovens

ROTARY TOASTING OVENS

Rotary toasting ovens have been the standard ovens for toasting all types of flakes since the inception of the breakfast cereal industry. Flakes are difficult to toast in a fixed position on flat bands; they tend to toast like a brown-edge cookie, with the edges browning while the center remains pale. Rotary ovens toss or suspend flakes in heated air, which heats all surfaces evenly and results in uniform browning and toasting.

One of the earliest rotary ovens was designed between 1897 and 1907. Shouldice Brothers was incorporated in 1907 and has manufactured these ovens since that time. No patents were applied for originally, but several patents have since been issued to others in the field (James, 1943; Schlotthauer, 1973).

Oven Construction

The rotary toasting oven consists of an insulated outer shell, an inner revolving cylinder, gas burners, a waste hopper with screw conveyor, and a drive mechanism. The oven is mounted on legs high enough to allow the floor underneath to be cleaned. A standard-size unit is generally 5 ft 2 in. wide, 8 ft 7 in. high, and 23 ft 8 in. long and weighs about 6,500 lb. The oven capacity in this size range is 500–700 lb/hr infeed of corn flakes at 13–15% moisture (400–600 lb/hr discharge of toasted flakes at 2.5–3% moisture).

The main body of the oven is double-shell construction. The interior shell is made of carbon steel, and the exterior shell is preferably stainless steel for ease of maintenance, with insulation sandwiched between to reduce heat loss.

The rotary cylinder may be constructed of either perforated or nonperforated steel. The perforated cylinder is used to toast larger flakes, such as corn, wheat, or rice flakes. The perforations are usually 3/32 in. in diameter on 1/16 × 7/64 in. centers and are staggered, with 1-in. borders at the ends. Usually 16-gauge cold-rolled steel is used for the cylinder. The nonperforated cylinder is used for toasting small products such as bran cereals, where the product might sift through perforations.

Oven Operation

A key aspect of the design of rotary toasting ovens is the ability to incline the cylinder from a level position to one where the infeed end is up to 10 in. higher than the discharge end. The pitch of the cylinder for each product being toasted is a critical control variable, but once established for a product, it need not be varied from day to day or hour to hour. The pitch of the cylinder is used in conjunction with its rotation speed to obtain the desired retention time.

The rotation speed is set so that if the cylinder is rotating clockwise when viewed from the discharge end, the product is cascading in a "waterfall" pattern beginning at the 11 o'clock position. If counterclockwise rotation is used, the waterfall should begin at the one o'clock position. The cylinder is driven by a variable-speed drive. Typical speeds are in the range of 13–17 rpm, and the average retention time for flakes is about 3 min.

The cylinder is arranged so that the moist flakes entering the oven fall directly over the perforations and therefore come in immediate and direct contact with the heated air. This is important in cases such as corn and rice flakes, where a blistering effect on the flake is desired.

Lifters

An important feature built into the design of the drums, whether they are perforated or solid steel, is the "lifters" on the inside surface. Their function during the rotation cycle is to lift the product to the 11 or one o'clock position so that "waterfalling" can take place. Without their assistance, at least some of the product would sit in the bottom quadrant of the cylinder and would not be suspended in the heated air. Typically there are nine equally spaced, triangular lifters, each 1.75 in. high.

Removing Fines

A V-shaped waste hopper—an integral part of the oven housing—is located under the toasting cylinder. It is insulated to prevent heat loss, as is the oven housing surrounding the cylinder. The walls of the waste hopper are steeply pitched to prevent fines from sticking to the walls and causing oven fires. The fines are continually removed from the waste hopper by a screw conveyor located in the bottom of the hopper.

Oven Burners

Rotary toasting ovens are heated by natural, manufactured, or propane gas. The burners are normally ribbon style (drilled pipes), as shown in Figure 5. Four are located inside the oven, two on each side of the drum. In this way the flame may be raised or lowered at either end of the oven to distribute the heat as desired. Burner controls are

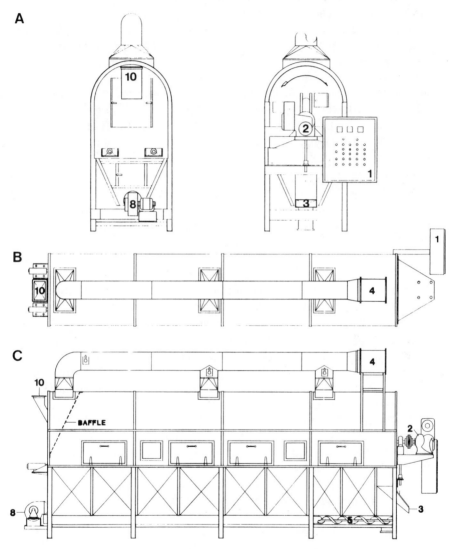

Figure 5. Rotary cereal toasting oven with ribbon-type gas burners. A, end elevation; B, plan view; C, side elevation. 1, control panel; 2, 1.5-hp variable-speed drive (3.2–32 rpm); 3, toasted flake discharge; 4, exhaust fan (2,000 cfm); 5, 1.0-hp, 68-rpm Syncrogear; 8, cast-iron blower (1,100 cfm); 10, untoasted flake feeder. (Courtesy Shouldice Bros. Co., Battle Creek, MI)

all at the discharge end of the oven so the flakes may be observed while oven heat is being adjusted. A fifth burner is used to preheat makeup air, which is injected into the chamber below the oven cylinder.

Another oven design, shown in Figure 6, employs two combustion air blowers instead of the five ribbon-style burners. One burner is mounted in a duct on the side of the feed end, positioned on the lower side of the oven on which the product is cascading or waterfalling, as shown in Figures 5 and 6. Having the heated air at the feed end hit the raw flake side of the cylinder first overcomes the cooling effect of the raw flakes at this point. The second burner is mounted in a duct on the lower side of the discharge end. This duct and burner are positioned on the opposite side from the feed end duct and burner, so that the burner does not overtoast the flakes.

Exhaust and Makeup Air

In the type of oven shown in Figure 6, a portion of the exhaust air from the exhaust duct is returned to each supply duct ahead of the burners. Makeup air also enters these supply ducts ahead of the burners instead of under the cylinder as in the ribbon setup. The ratio of exhaust to supply air is about 50:50. Because the makeup air doubles in volume once it is heated by the burner, care must be taken not to run too much of it. If too much is used, the pressure inside the oven becomes positive, whereas a slight negative oven pressure is desired for good operation and exhaust of moisture.

Three exhaust ducts are located on the top of the oven, with dampers to control the airflow of each. The damper over the feed end is usually run wide open, and the other two are dampered down. The three ducts are joined into one containing the exhaust fan. The ribbon burner model has a damper in the makeup air duct as well as a fresh air makeup fan. A baffle surrounding the cylinder in the first third of the feed end of the oven is positioned in such a way that the air being exhausted in the first exhaust duct must be drawn through the cylinder at the feed end. This arrangement draws off the initial high moisture load coming from the product. For best drying and toasting, the airflow through the cylinder should be 250–300 ft^3/min (cfm).

Toasting Temperatures

Bimetal temperature sensors are mounted in the exhaust ducts and in the lower chamber and are connected to Partlow-type temperature controllers. Wheat and oat cereals can generally be toasted at around 350–450°F (177–232°C). Rice and corn flakes require higher toasting temperatures, generally in the 450–600°F (232–315°C) range. For most applications, the feed end of the oven is run at a higher temperature than the discharge end.

Cleaning and Safety Considerations

The oven cylinder picks up small fragments of product over extended usage. These pieces char and blacken and either stick in the perforations or adhere to the cylinder itself around the lifters. When such deposits build up, they break off and contaminate the product flow with black specks.

Star-shaped metal pieces 0.5-0.75 in. across, obtained from foundry molding processes, can be used to abrade these deposits off the interior of the cylinder. About 1-2 ft^3 of the stars are fed into the oven while the cylinder is rotating and are collected at the discharge end. This is repeated until little or no black material comes out with the stars.

Rarely must an oven cylinder be removed from the oven housing for cleaning. This occurs only when perforations, mostly at the feed end, become so clogged that they cannot be cleaned in place. This clogged condition greatly reduces airflow through the cylinder and causes high-moisture product and poorly developed or blistered flakes.

Quick-release doors in the oven housings release pressure in case of an explosion inside the oven or duct work. The explosion door area is usually 1 ft^2 for each 15 ft^3 of oven and duct work.

CONVEYORIZED (FLAT-BAND) OVENS

In the preceding section, we described how cereal flakes are toasted by being tossed or suspended in hot air in a rotary oven. Because they are small and thin, the flakes would easily burn on a flat band or conveyor. Large products, on the other hand, or those that are 0.25-0.75 in. thick, would bend, twist, and/or crumble if they were toasted by being tossed in an airstream. These larger products, such as compressed-flake and shredded-grain biscuits, are suitable for baking in a fixed position on a flat surface.

Conveyorized or band ovens are quite simply ovens in which a conveyor moves through the oven chambers from one end to the other. The unbaked, shaped product is placed on the band at one end, and the finished, baked product emerges from the other end. The shredded-grain biscuits typically baked in band ovens may be made of wheat, corn, rice, or oats; wheat biscuits are predominant. As described in Chapter 2, these vary in the number of layers of shreds from two to 20.

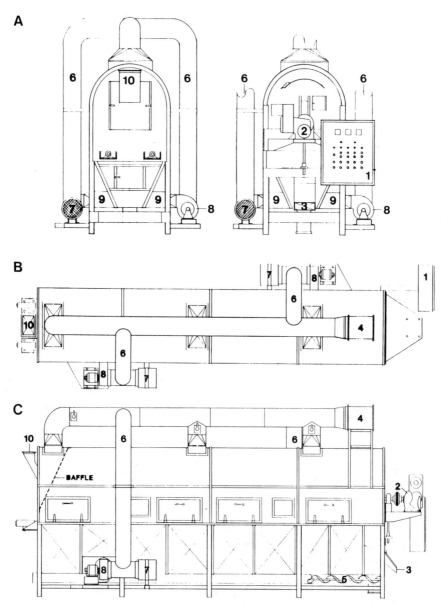

Figure 6. Rotary cereal toasting oven with forced-air heating. A, end elevation; B, plan view; C, side elevation. 1, control panel; 2, 1.5-hp variable-speed drive (3.2–32 rpm); 3, toasted flake discharge; 4, exhaust fan (2,000 cfm); 5, 1.0-hp, 68-rpm Syncrogear; 6, return air (600 cfm); 7, fresh air (500 cfm) with filter; 8, cast-iron blower (1,100 cfm); 9, burner box; 10, untoasted flake feed. (Courtesy Shouldice Bros. Co., Battle Creek, MI)

Direct or Indirect Heating

Conveyorized ovens can be either direct- or indirect-fired. In direct-fired ovens, burners are mounted inside the oven baking chamber, and the product is thus exposed directly to the heat source, such as gas burners or electric heating elements. Direct-fired band ovens to our knowledge are not used in the manufacture of breakfast cereals. After World War II, when band ovens became the preferred and most common type of oven used in the industry, indirect-fired ovens were found to be more efficient, and products baked in them were more uniform in quality (Gales and Perry, 1960).

In indirect-fired band ovens, the burner is mounted close to a fan that circulates heated air (and combustion products if gas or oil is used as fuel) to the baking chamber and onto the product to be baked. These heated gases blowing on the product effect the heat transfer and moisture removal that constitute the baking and toasting process. A typical indirect-fired, forced-convection oven used for cereal drying and toasting is shown in Figure 7.

Figure 7. Typical indirect-fired forced-convection oven for toasting cereals. The oven is composed of 10 sections or zones, which are identical except for the infeed and discharge zones. The operator in the photograph is standing at the control panel for one of the zones. The dark object directly over his head is the blower and control package of the gas burner for that zone. At the extreme right on the side of the oven is a small dial, which indicates how long the band takes to travel through the oven. The dial is mounted on the control for adjusting bake time. (Courtesy Spooner Oven Co.)

Oven Operation

Figure 8 shows side and end view cross sections of a forced-air convection oven. The conveyor on which the product rests is represented by the horizontal line 1. Above and below the conveyor and extending across the full width of the oven are four pressure chambers (labeled 2-5 in Figure 8). The pressure chambers are supplied with hot air from a double-outlet circulating fan through two ducts. The surfaces of the pressure chambers are fitted with a series of nozzles, slots, or tubes. The air enters the baking chamber and is directed onto the product at high velocity through these devices. It gives up a portion of its heat to the cereal being carried through the chambers and then escapes laterally back to the fan inlet for recirculation. The heat loss in the circulating air is made up by heat from the burner. In ovens fired by gas or fuel oil, the burner tube is positioned under the fan suction inlet, and products of combustion are mixed with the circulating air before it enters the fan, as shown in the end view cross section in Figure 8.

The air temperature in the duct on the discharge side of the fan is used to control the firing of the burner. The bake may be further regulated by varying the velocity of the air issuing from the nozzles of any one of the four pressure chambers. These velocities are controlled by dampers (labeled 12-15 in Figure 8). An exhaust outlet is provided at a convenient point in the zone to draw out air equivalent in volume to the new products of combustion entering the baking chamber from the burner, together with any water vapor given off from the cereal being baked. The amount of exhaust is controlled by a damper in the exhaust stack and can also be augmented by an extraction fan.

Size and Output

The previous paragraphs explain how one oven section operates. In practice, multiple sections are joined together to form a completed oven. The overall length of the final oven depends on the product and the desired output at a given bake time; the longer the oven and the faster the band moves for a given bake time, the higher the output. Ovens may comprise as many as 10-12 zones or sections, each with its own burner and controls as just described. Zones vary from 13 ft 7 in. to 50 ft long and from 12 to 140 in. wide, depending on the desired output and the available space.

Design Features

Forced-air convection ovens are very efficient. They have high moisture removal rates because of the convective air movement, plus

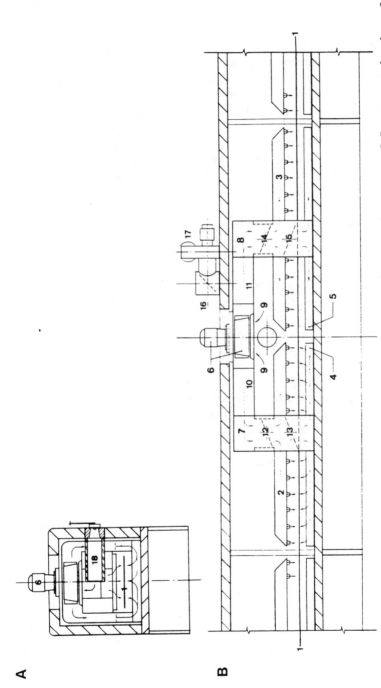

Figure 8. End (A) and side (B) view cross sections of a forced-convection oven. 1, conveyor; 2–5, pressure chambers; 6, double-outlet circulating fan; 7 and 8, ducts; 9, fan inlet; 10 and 11, delivery ducting; 12–15, dampers; 16, exhaust outlet; 17, exhaust fan; 18, burner.

naturally occurring radiant heating from the oven surfaces that contributes the unique flavor development available only from such heat sources.

One of the most important features of these ovens is the design and operation of the nozzles or openings in the pressure chambers through which the heated air is directed onto the product. Holes were the first design used. Then came slots across the whole width of the chamber. Tubes extending down (or up) from the pressure chamber are the most recent design. These tubes normally terminate only 0.5 in. from the product and are spaced 2–2.5 in. apart. They are much more efficient than holes or slots but if not properly designed and operated, they can overdry the product before a good toasted flavor has developed. To avoid the overdrying effect, the air velocity is usually kept in the range of 3,000–4,000 cfm or lower.

The design of the pressure chambers has also changed over the years. Initial styles did not extend the full width of the band, which meant that the nozzle sections overhung the pressure chambers on each side to cover the full width of the band. This design allowed plenty of room on the sides of the chambers for fast return of air to the burners. In the new designs, the pressure chambers extend the full width of the oven band. The extra radiant baking effect that comes from the larger surface of the pressure chambers more than compensates for the slower air return in these models.

Exhaust Air and Humidity

The amount of air exhausted from the oven has a very significant effect on the baking process. This exhaust naturally varies with oven loading and type of product but usually runs 700–1,500 cfm; 1,500 cfm is the maximum for almost all products.

Baking of shredded wheat and oat products is more efficient if the oven exhausts are kept almost closed in most of the first half of the oven sections. This builds humidity, and humid air carries more heat to the product than dry air. In addition, because the surface of the biscuits does not dry out as fast in the humid air, the biscuits rise more and moisture removal is faster; case hardening, or drying of the surface of the piece being baked, impedes the removal of moisture from the center.

With corn or rice products, the strands must become puffed during baking. To achieve this on a product entering the oven at, say, 30% moisture, the moisture content must be lowered to 11% in 30–40 sec using air at about 300°F (150°C). The product must then be fluidized in much hotter air (550–600°F [288–315°C]) to bring about the puffing action.

Oven Control

The dampers that control the oven air velocity are very important regulators of the baking process. As mentioned above, air velocity above 4,000 cfm, particularly in the early part of the oven, dries most shredded products before a pleasant toasted flavor can develop. If the product dries and the color develops too late in the oven, the tops of the biscuits will be weak. Excessive velocity can also cause burned edges on most products. Air velocity also influences the expansion of the product; higher velocity usually promotes expansion.

In summary, only four variables control the forced-convection baking process: temperature, air velocity, bake time, and amount of exhaust. Controlling the bake is not a simple matter, however. A 10-zone oven, for example, has 10 temperature controls, 40 velocity dampers, one bake time variable, and 10 exhaust dampers. Looking at it this way, one can see the advantages of modern electronic programmable logic control (PLC) equipment!

Mechanical Details

The natural gas or propane burners most used in forced-convection ovens are prepackaged and have a 45:1 turndown ratio. Oven bands are typically wire mesh; solid and perforated bands are generally not used for cereal products. The openness of the wire mesh allows good airflow and moisture removal from the product. Almost all wire mesh bands are made from carbon steel; stainless steel is unnecessary because the heat of the oven prevents rusting of the carbon steel. Stainless steel is necessary in ovens with clean-in-place (CIP) systems because the water added by such systems may not all be evaporated by the heat that remains in the oven after shutdown.

Controls for regulating temperature are standard types used in the baking industry. Today most firms want to be able to tie in these oven controls to a PLC system for automatic operation of the oven. Most oven manufacturers now offer electronic control systems.

Sanitation and Safety

Forced-convection ovens should be designed for easy cleaning, with cleanout doors on the sides. Crumbs that fall off the product and through the wire mesh band land on the bottom pressure chamber and can be brushed or vacuumed off. Some nozzle systems are built in a ladder design that allows crumbs to fall below the bottom pressure chamber onto a deck, from which they are then easily removed. Cereal ovens normally require cleaning only once a week. If an oven is cleaned less often, or if crumbs are allowed to build up, the crumbs may ignite and cause oven fires.

CIP systems are installed in some cereal ovens, but they are expensive. They may be needed to avoid corrosion in ovens where vitamin-fortified products are baked.

Over time, exhaust stacks accumulate bits of carbon from product fines carried from the oven by the exhausted air. The biggest fire hazard in exhaust stacks occurs when no product is flowing through the oven but temperatures have not been reduced. Without the cooling effect of the water vapor being carried out of the oven during baking, stack temperatures can rise rapidly and cause the accumulated fines to ignite. Stacks should be cleaned when inspections show they need it.

The safety features on burners are beyond the scope of this discussion and are typically controlled by government and insurance regulations. The oven housings are built with explosion panels located primarily in the oven top. Should an explosion occur, its force would be directed to these "soft" areas in the oven roof. Usually there is about 1 ft^2 of "soft" area for every 15 ft^2 of oven area.

Before gas-fired ovens can be lighted, the oven must go through a "purge" cycle, usually lasting 5 min. During this time there are about four complete changes of air within the oven. Purging ensures that no gas accumulates in the oven before the burners are fired. During the purge cycle, all exhaust dampers are fully open regardless of where they are normally set during baking.

AIR-IMPINGEMENT TOASTING

Like the traditional rotary toasting oven already discussed, the high-heat-transfer air-impingement convection oven can suspend products during toasting. Jetzone is the registered trademark of Wolverine Corporation (Merrimac, MA) for such processors. Similar equipment is available from others. This forced-convection process is based on the jet-tube principle: air is forced through jet-tube nozzles, which direct high-velocity jets of heated air into the product treatment area, surrounding the product with air. The result is uniform heat transfer and uniformly treated product at relatively short dwell times.

Jetzone dryers normally have multiple zones, each with adjustable velocity and temperature controls for operating flexibility (Figure 9). They are used on raw or intermediate grain products, including cereal pellets, flakes, and shapes. Like traditional rotary ovens, they are normally not suitable for toasting large shredded-grain biscuits or biscuits made of compressed flakes, which are better toasted in more conventional conveyorized flat-band baking ovens.

How It Works

In the fluid-bed process, air directed from jet tubes impinges on a nonperforated belt or pan conveyor. The air reflects off the solid surface, literally lifting and surrounding the product. Each particle is enveloped by air at a sustained temperature of up to 600°F (315°C); the product "floats" in the treatment zone.

Jet tubes range in diameter from 0.125 to 1.75 in. depending on the bulk density and the fragility of the product. Steels used in the units also vary depending on the application. Pans may vary in width from 9 to 50 in., depending on how much product flows through the dryer.

Treatment time can be varied by changing both the number and length of 10-, 20-, or 30-ft sections, since oven length controls dwell time. Separate zones (one, two, or three) are used to control and direct the airflow around the product.

Air-impingement dryers normally use recirculated air. They can be run, however, from 100% recirculation air down to a one-pass system, or 100% exhausted air. The normal recirculation level for cereal processes is 85–95%, with makeup air replacing the amount of air exhausted.

A system of three dampers per zone is used to control the airflows. A single circulating air damper controls the amount of air going to

Figure 9. Jetzone air-impingement dryer with two zones. (Courtesy Wolverine Corporation, Merrimac, MA)

the jet tubes. The other two dampers control the amounts of recirculated and exhausted air. Humidity is not generally added to the airstream. Rather, relative humidity within the oven is controlled by exhaust air adjustment. Damper controls can be either manually adjusted or motorized.

Thermocouples for sensing temperatures are mounted in the pressure plenum over the jet tubes. In this way they measure the temperature of treatment air rather than of recirculation air. Air velocities range from 2,000 to 12,000 ft/min, measured at the jet-tube tip.

Air-impingement dryers can be operated by total manual control, manually from a main control panel, or totally by PLC. The control functions are burner ignition, temperature of each zone, velocity of air through the jet tubes, dwell time of the product (conveyor speed), makeup air damper setting, exhaust air damper settings, and product feed rate to the oven.

Depending on space requirements and product characteristics, units are available in a number of configurations. All fans and burners may be integral with the unit, with collectors remote, or collectors, burners, and fans may be separated from the unit. Either an oscillating conveyor or a nonperforated belt conveyor may be used.

Typical Cereal Line

A typical cereal line using this type of oven-dryer consists of one or two dryers and a cooler. A two-unit line includes a dryer and a cooler, and a three-unit line has two identical dryers in line and a cooler. Each cereal dryer typically has two treatment zones, while the cooler has only one (Figure 10).

The raw or half-product from an extruder, flaking mill, or other intermediate process is conveyed to the dryer. As with other conveyor-ized oven dryers, the first zone in any line uses extremely high temperatures to remove moisture. This also has a puffing action. Puffing temperatures range from 400 to 500°F (200–260°C). As the product cooks, moisture is reduced and temperature in the next zone(s) must be lowered to prevent burning while the product continues cooking. This is the toasting part of the process. It further dries the partially dried, puffed product to supply the specified color and crunch. Toasting temperatures range from 300 to 400°F (150–200°C).

In a two-unit line, the first treatment zone acts as the "puffer" and the second zone performs the toasting. A single zone is sufficient for puffing and toasting, but using two dryer units in line enables a faster flow for higher production. In a three-unit line, the first dryer in line is designated the puffer; the initial high-temperature zone removes moisture, and the second zone begins the toasting process. The second dryer is designated the toaster and continues the toasting begun in the second puffer zone.

If cereal grains are not cooled immediately after toasting, they continue to cook and may burn. The cooler, the last unit in both two-unit and three-unit lines, lowers the product temperature to stop the cooking process and prepare the product for further processing or packaging.

Safety and Sanitation

Safety standards for air-impingement ovens are normally those required by government or insurance regulations in the United States. Explosion doors are sized to conform to the volumes of air in the units and are equipped with explosion relief latches on the access doors. Should an explosion occur, the doors would open to prevent internal damage to the unit.

Figure 10. Jetzone air-impingement dryer with three zones, two for puffing and toasting and one for cooling. (Courtesy Wolverine Corporation, Merrimac, MA)

Product changeover and cleanup are simple because these ovens have a minimum number of moving parts, and all interior areas are 100% accessible for cleaning. CIP systems are available and can be controlled either manually or by PLC.

OPERATOR RESPONSIBILITIES

The oven operator has one of the most important positions on a cereal processing line. The operator is responsible for the color, moisture content, and bulk density of the finished product and, in the case of a flaked cereal, must work closely with the flaking roll operator to ensure that the flakes are the proper thickness. Flakes are checked manually against known, standard quality products, and bulk density and moisture are measured frequently. Color is also checked against quality standards. As pointed out in Chapter 2, instrumentation is now available to measure and control some of these variables on-line. Nevertheless, as the last operator in the processing line, the oven operator may be the first to know when finished product quality is substandard.

References

Gales, D. R., and Perry, M. H. 1960. Oven developments—The forced convection oven. Food Trade Rev. 30(3):56.

James, T. R. 1943. Toasting flaked cereal and the like. U.S. patent 2,317,532.

Jankowski, T., and Rha, C. 1986. Retrogradation of starch in cooked wheat. Starch/Staerke 38:6.

Lauhoff, F. 1913. Roll. U.S. patent 1,062,170.

Schlotthauer, W. W. 1973. Cereal toasting oven. U.S. patent 3,745,909.

White, E. G. 1979. Process for manufacturing a whole wheat food product. U.S. patent 4,179,527.

Chapter 6

Unit Operations and Equipment IV. Extrusion and Extruders

ROBERT C. MILLER

In Chapter 3, extrusion was introduced as a continuous cooking process. In this chapter, the extrusion process is explored in more detail, leading to two important specific applications in the breakfast cereal industry—direct expansion and the forming of solid shapes. These date to the origin of food extrusion and, in the case of forming, to the earliest uses of the extruder. The discussion is limited to screw extrusion of three major types: single-screw and intermeshing twin-screw, the latter in counterrotating and corotating styles.

Extruder Components

Extruders consist of two basic components (which may be further subdivided into process areas): screw(s) rotating in a barrel and propelling the feed material forward while generating pressure and shear, and a die or restrictive orifice through which the product is forced. These components interact to create the desired process conditions, but they are best treated as separate entities for analysis. When separated, the die "sees" the entire preceding process simply as a pressurized fluid, and the screw and barrel section "sees" the die merely as a resistance.

For any particular extruder and product, the behavior of each component can be characterized by its response to changes in operating pressure—the pressure of the flowing product as it emerges from the screws. Intuitively, one can see that increasing pressure forces more product through the resisting die, as shown in the die "operating line" in Figure 1. This line passes through the origin (no pressure = no flow) and indicates that flow rate increases with pressure.

An analogous operating line may also be constructed for the screw component, which reacts to pressure in the opposite direction: flow is reduced at higher operating pressure. The flow rate is highest at zero pressure ("open discharge," or no die resistance) and falls to zero at a characteristic high pressure ("closed discharge," or infinite resistance). Operation of the combined system is defined by a single point—where the two lines intersect.

The positions of the operating lines for any particular application depend on screw and die geometry, among other things. As these are changed, the lines and operating points shift, as indicated in Figure 2. With smaller dies, for example, the die resistance may be increased, leading to a smaller flow rate for a given pressure. Similarly, screw speed and shape strongly influence the open discharge flow and elasticity of response to pressure. Regardless of where the operating lines are located, however, the operation of the system may be found at their unique intersection.

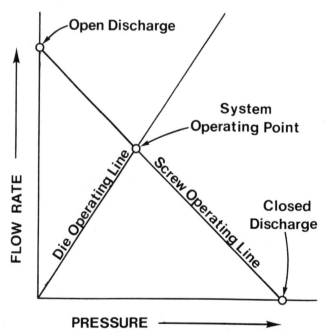

Figure 1. Simplified die and screw operating lines, showing extruder operating point at their intersection.

Figures 1 and 2 represent an ideal situation. In real processes, the lines are seldom straight, due to nonideal rheology and to other factors, such as temperature, that affect viscosity. This is especially true in cooking extrusion, where rheological properties can change dramatically during the process and in response to changes in operating pressure. Ideal extruder behavior is most closely approached in forming processes, in which the product is already cooked and temperature changes are usually minimized. In either case, the operating line principle is valid, except not necessarily in a simple linear fashion.

Principles of Die Flow

The die operating line, a function of geometry and rheology (Rossen and Miller, 1973), is expressed as:

$$Q_{die} = kP/\mu , \tag{1}$$

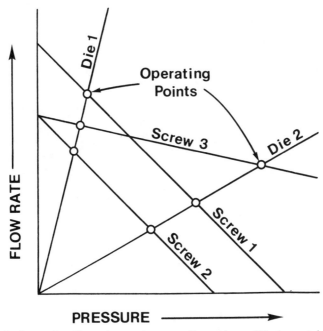

Figure 2. Operating lines for various configurations. Die 1, restrictive (small die constant); die 2, less restrictive; screw 1, elastic pressure response; screw 2, decreased capacity and pressure capability (reduced screw speed can cause this); screw 3, stiff (less elastic) pressure (shallow flights or twin-screw extrusion show this kind of response).

where Q_{die} = volumetric flow rate, P = operating pressure, μ = product viscosity, and k = the die constant.

From this, it can be seen that the line moves upward from the origin with a slope equal to the die constant divided by the *viscosity*, so that for any particular die geometry constant k, the slope is inversely proportional to viscosity. The concept of viscosity implies an ideal Newtonian fluid in which the resistance to fluid flow (shear) is constant. This is seldom encountered in cereals, which typically exhibit both *yield stress* (some minimum pressure must be applied before flow begins) and *shear thinning* (apparent viscosity decreases at higher flow rates). These properties cause the operating line to start at a point to the right of the origin and then to proceed upward in a concave curve, even at constant temperature (Figure 3). These phenomena are subjects of current research and are discussed further in the literature. In practice, the operating line is best determined experimentally. As a first approximation, and in many cases throughout the process analysis, the idealized linear model is sufficient.

The other major product variable is *temperature*. Except in those truly isothermal systems that are approached in forming, product temperature varies with operating conditions, with corresponding

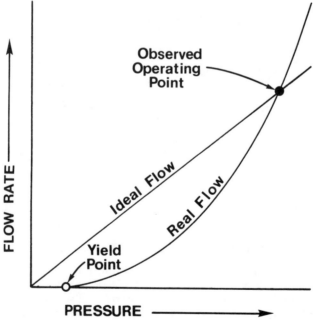

Figure 3. Actual and ideal die operating lines for cereal extrusion (schematic).

variations in viscosity. At the die, the viscosity generally decreases with increasing temperature. By heat transfer at the die, this effect may be controlled or put to good use in forming.

Die geometry, represented by k in equation 1, is another component of flow resistance. For any particular viscosity, the resistance is proportional to $1/k$. Generally, this resistance term is greater for dies that are longer or have smaller cross-sectional dimensions. Die constant equations for various simple geometries, assuming ideal behavior, have been derived from the equation of motion and are shown in Figure 4. Resistance increases with length but increases much more rapidly with decreases in cross-section dimensions; "slot" die resistance, for example, is inversely proportional to the cube of slot thickness. For this reason, the resistance of passages upstream from the final smallest dimension may usually be ignored—almost all of the pressure drop in the die occurs in the most constricted segment. Where a more rigorous approach is desired, the resistances of other locations in the flow path may be added to that of the final die dimensions:

$$1/k = 1/k_1 + 1/k_2 + ... + 1/k_n , \qquad (2)$$

where $1/k_1$ = total die resistance and $1/k_i$ = resistance of each portion of the flow path in sequence.

Cereal processing normally requires multiple die openings. These resistances are in parallel, so that the overall resistance seen by the extruder is found by adding together the individual values of k. When the die holes are all identical, the total die constant is the product of individual k times the number of orifices. In practice, the individual k values should be very similar, even when a variety of shapes is used, to assure flow uniformity—flow tends to "short circuit" to the largest hole. Uneven pressure distribution due to particular extruder geometry may cause uneven flow, even through identical dies. In this case, the flow may be equalized by adjusting individual resistances to compensate for the variation, usually by changing die land length (i.e., the length of the final, most restrictive portion of the die).

Entrance losses or additional pressure drops due to changes in the cross-sectional shape of the flow path may also be calculated from equations in the literature (Padmanabhan and Bhattacharya, 1989), but these are usually very small at the high viscosities and low velocities encountered in extrusion (Wilkinson, 1960).

Entrance geometry influences the flow in another more significant way, however, due to another property of cereal flow—*elasticity*. If the entrance to the final die orifice changes abruptly from a larger cross

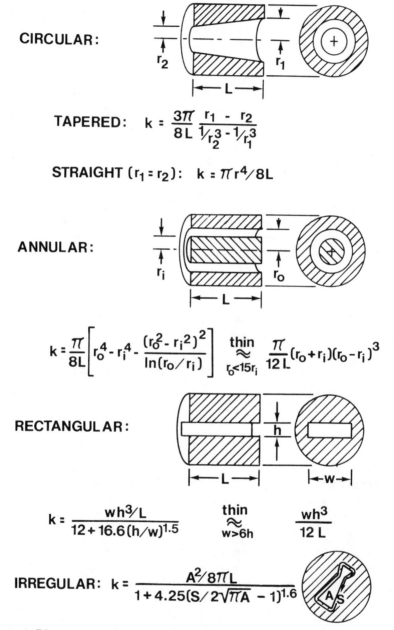

CIRCULAR:

TAPERED: $k = \dfrac{3\pi}{8L}\dfrac{r_1 - r_2}{1/r_2^3 - 1/r_1^3}$

STRAIGHT $(r_1 = r_2)$: $k = \pi r^4/8L$

ANNULAR:

$$k = \dfrac{\pi}{8L}\left[r_o^4 - r_i^4 - \dfrac{(r_o^2 - r_i^2)^2}{\ln(r_o/r_i)}\right] \quad \underset{r_o<15r_i}{\overset{\text{thin}}{\approx}} \quad \dfrac{\pi}{12L}(r_o+r_i)(r_o-r_i)^3$$

RECTANGULAR:

$$k = \dfrac{wh^3/L}{12 + 16.6\,(h/w)^{1.5}} \quad \underset{w>6h}{\overset{\text{thin}}{\approx}} \quad \dfrac{wh^3}{12L}$$

IRREGULAR: $k = \dfrac{A^2/8\pi L}{1 + 4.25(S/2\sqrt{\pi A} - 1)^{1.6}}$

Figure 4. Die constants for various geometries. Equations for thick rectangle and irregular shapes are semiempirical. Other are exact for ideal fluids. A = cross-sectional area; S = length of periphery.

section, and the die length is short enough so that the fluid stresses have insufficient time to equilibrate, the emerging product tends to rebound elastically from the more compressed state and to enlarge in cross section from the die dimensions. Longer, more streamlined dies produce less of this rebound effect (Harper, 1986; Hawkins and Morgan, 1986).

Another elasticity effect is caused by the velocity distribution in the die. In the equations given, fully developed laminar flow is assumed—and is usually the actual case. Physically, this means velocity at the die surfaces is zero and increases smoothly to a maximum value at a central point in the flow path. With this velocity profile, product layers slide past one another in shear, causing entanglement and stretching of long molecules (particularly starch), which are responsible for the elastic nature of cereals. Upon emergence from the die, the velocity profile becomes flat in the absence of applied shear—the slow parts speed up and the fast parts slow down so that they match. The stretched molecules tend to rebound at this point, forcing the emerging stream to enlarge.

In die-face-cut products, still another result of elasticity is observed—product rounding. In a nonelastic fluid, cut pellets are thicker in the middle than at the edges, reflecting the velocity distribution in the die. As each pellet emerges from the die, it is cut off at the die surface, leaving a rounded front surface and a flat rear surface. With an elastic fluid, however, the internal stresses operate after the cutoff to make the rear surface rebound into a convex shape, producing a pellet with more symmetry (Miller, 1985).

All of these elastic effects depend on velocity distribution, which requires that the product adhere to the die surfaces. For this reason, most extrusion dies are not made with overly smooth inner surfaces. When the elastic effects are not desired, however, the velocity distribution may be reduced by using smooth inner surfaces, such as Teflon linings, to cause the product to slip through the die instead of flowing in a normal fashion (Donnelly, 1982).

Elastic swelling of the product must not be confused with product expansion—they are independent phenomena, although they often occur together. Nonexpanded products may swell. Product expansion, on the other hand, is caused by steam and will be discussed in more detail later.

Principles of Screw Flow

We now turn to the pressure-generating portion of the extruder, the screw and barrel. Of the machine styles used for cereal extrusion, three very different flow mechanisms are encountered. These are considered separately and in comparison.

COUNTERROTATING TWIN-SCREW EXTRUSION

Despite its mechanical complexity, the counterrotating twin-screw extruder has the simplest pumping mechanism of the three styles and is therefore the best with which to begin discussion of principles. As shown in Figure 5, the closely intermeshing screws create a series of discrete chambers along each screw. Each chamber consists of a segment of screw channel, separated from other segments by the interposed tips of the intermeshing adjacent screw flights. As the screws rotate, the chambers progress toward the discharge, conveying the product by positive displacement. The volumetric flow rate is therefore inelastic (independent of operating pressure) and is calculated as the product of the number of chambers passing any point per unit time and the volume of each chamber:

$$Q = 2V_c nN \,, \tag{3}$$

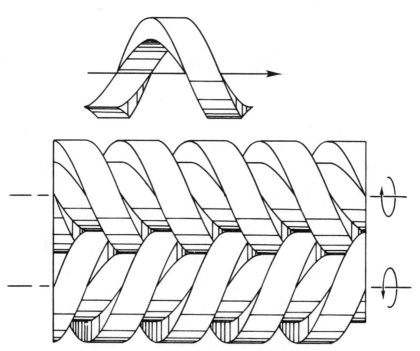

Figure 5. Counterrotating twin-screw extruder. Bottom, intermeshing screws; top, chamber formed in screw channel by adjacent screw flight tips and barrel surface.

where Q = volumetric flow rate, V_c = chamber volume (measured or calculated from screw dimensions), n = number of parallel leads (flights) on each screw ($n = 2$ for commonly used twin-lead screws; for multiple-lead screws, more than one chamber passes by with each revolution), N = screw speed (revolutions/time), and the factor 2 is included because the two screws operate in parallel.

Since all real machines must have tolerances between moving parts, this ideal equation must be corrected for leakage—a small backward flow of product through gaps between the screws and between the screws and barrel:

$$Q_{net} = Q - Q_l , \qquad (4)$$

where Q_{net} = actual flow and Q_l = leakage flow.

Leakage flow is described in detail by Janssen (1978) as a function of geometry and pressure (more on this later).

A constant volumetric flow rate is not practical for compressible products for which productivity is limited by low feed density. For these products (such as low-density flour blends), tapered screws with large-capacity feed sections are available.

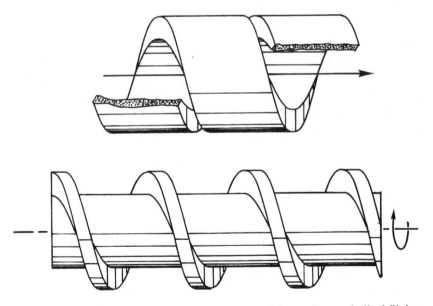

Figure 6. Single-screw extruder. Bottom, screw with continuous helical flights; top, channel formed by screw flights and barrel surface—the channel is an open-flow path throughout the extruder length.

SINGLE-SCREW EXTRUSION

Mechanically simple, the single-screw extruder employs a more complex flow mechanism than that of the counterrotating twin. Whereas the twin is essentially a positive displacement device with a small correction for leakage, the single-screw extruder leaks freely throughout the length of its open helical screw channel (Figure 6). How, then, can the single-screw machine generate pressure?

If the product were a solid, the normal mechanical advantages of screw motion would apply—provided that the product slips on the screw surfaces and adheres to the barrel surfaces in rotation but not axially, in corkscrew fashion. Indeed, this situation is encountered in some cereal processing operations, particularly with low-moisture products, and in feed sections where the material has not yet developed fluid properties.

In most cases, however, the product does act like a fluid in that it 1) flows by yielding and 2) adheres to all surfaces (with exceptions noted later). By analogy to the twin-screw mechanism, it is useful to divide the flow into two components for better understanding. These are a forward or "drag flow" component and a backward or "pressure flow" component; the latter is similar to the leakage flow of the twin, but of far greater significance.

Drag Flow

Drag flow is caused by the relative motion of the screw and barrel surfaces. In looking at the counterrotating twin-screw extruder, we imagined moving along the rotating screw with the progressing chamber. If we examine the single-screw extruder from the same point of view, we see that the flights again move downstream, but at the same time, product is dragged back upstream by the screw surfaces, reducing the net forward flow. Although the extruder might be amenable to analysis from this perspective, it is easier to understand after a change in point of view—to that of a stationary screw within a rotating barrel. At the velocities and viscosities encountered in extrusion, the two situations are mathematically identical.

As indicated schematically in Figure 7, the "rotating" barrel imparts a tangential motion to the product. This motion is redirected by the screw flights into the down-channel direction and is analyzed by dividing the original barrel velocity into two components, down-channel and cross-channel, at right angles to each other. The down-channel component is the main drag-flow term; the cross-channel portion causes circulation of the product as it encounters and is turned under by the "pushing" flight. The total result is that the product rotates as it proceeds down the channel—an important factor in heat exchange.

Since we assume that the product adheres to all surfaces, it is in a state of shear—product at the barrel surface moves downstream with it, while product at the screw surface remains behind. As shown in Figure 8, this results in a velocity profile, with the average product velocity somewhere between these extremes. In the simplest model, it is one-half of the downstream component of tangential barrel velocity. In slipping (solid) plug flow, the velocity is constant and equal to this barrel velocity component, with a resulting flow rate roughly equal to that of the ideal counterrotating twin-screw extruder.

Drag flow is calculated by applying the average downstream velocity over the screw channel cross section. For any particular screw design, this is independent of viscosity and is a simple function of screw speed and geometry:

$$Q_d = \alpha N, \tag{5}$$

where Q_d = volumetric drag flow and α = drag flow geometry parameter.

Screw Geometry

The geometry parameter, however, is not so simple. In general, it increases with screw diameter and channel depth, both of which

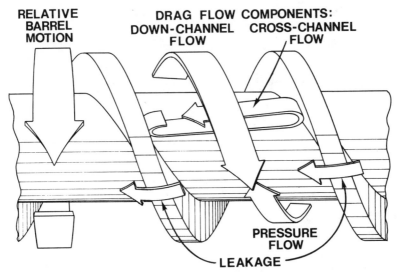

Figure 7. Flow components in the single-screw extruder.

increase the screw displacement. It also increases with pitch (the distance between flights, which also increases channel size), but only up to a maximum value, after which it declines with further increase in pitch as the cross-channel component of tangential velocity becomes larger than the down-channel component. More of the rotary motion is then converted to circulation and less to flow. In the extreme case

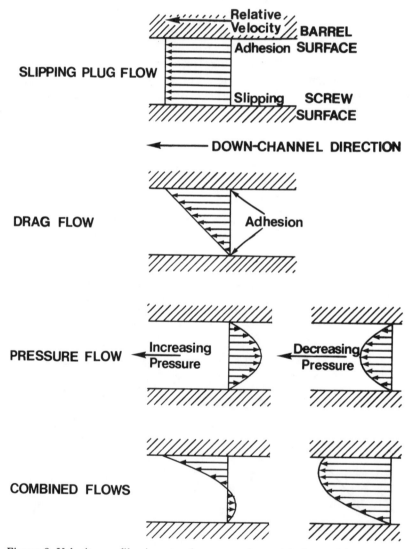

Figure 8. Velocity profiles in extruder screw channel (schematic).

of infinite pitch, where the flights are parallel with the screw axis, the extruder becomes a scraped-surface heat exchanger with 100% circulation and no forward flow. For simple screw designs with rectangular channels and shallow flights, the geometry factor is:

$$\alpha = \frac{1}{2}\,\pi^2 D^2 h \left(1 - \frac{ne}{t}\right) \sin\,\phi\,\cos\,\phi\,, \tag{6}$$

where D = screw diameter, h = channel depth, e = flight thickness (measured axially), t = screw pitch (measured axially), and ϕ = screw helix angle = $\tan^{-1}(t/D)$. These geometrical parameters are further described in Figure 9.

More rigorous equations for the effects of screw geometry, including deeper channels and temperature and viscosity gradients, are available in the literature (Bernhardt, 1967; Tadmor and Klein, 1970). However, these are of limited use for food extrusion since they all assume that rheological properties are known, which is the exception in food processing. Moreover, food extruders have been developed empirically, leading to effective but complex shapes that defy precise mathematical description. These include spiral and straight grooved barrels, cut-flight screws with unusual flight shapes, and interrupted flights and stators. Nevertheless, basic flow principles may be described qualitatively.

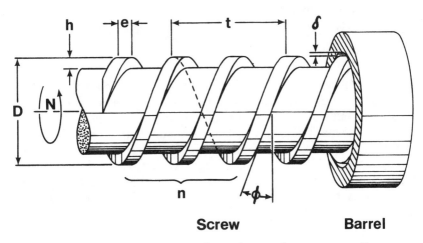

Screw   **Barrel**

Figure 9. Important extruder screw dimensions and parameters. D = screw diameter, h = channel depth, t = pitch, e = axial flight tip thickness, δ = clearance between flight tips and barrel, ϕ = screw helix angle (a function of diameter and pitch), N = screw speed (rpm), n = number of parallel leads or screw threads (in the diagram, $n = 2$).

Temperature Distribution

Temperature distribution in the screw channel, an important factor in heat transfer processes, can significantly affect drag flow. In the general case, where viscosity decreases with increasing temperature, barrel cooling and/or screw heating create a viscosity gradient, with the stiffest material at the barrel surface and more fluid material near the screw. This allows the screw to rotate more freely, approaching the condition of slip found in solid plug flow and increasing the drag-flow component. Conversely, barrel heating and/or screw cooling reduces drag flow by allowing slip at the barrel surface. Indeed, with total slippage at the barrel surface, the product merely rotates or "orbits" with the screw, producing no down-channel movement.

Even without an imposed temperature gradient, slippage at the barrel surface can be a problem when cohesion within the product, for example, is greater than adhesion between barrel and product. This is commonly addressed with grooved barrels, which provide more "grip" on the product. Grooves also permit backward leakage, however, mitigating their effectiveness in some cases. Spiral grooves can prevent slip while actually adding to forward impetus if they are pitched in the right direction (acting as an everted screw), or they can add to the mixing action if they are oppositely pitched. In some extruders, spiral groove depth varies along the barrel in opposition to a change in screw channel depth, causing a transfer of product from one flighted surface to the other. This adds a grinding or cutting mechanism to the process, which is particularly useful for low-moisture extrusion of coarse feeds.

Pressure Flow

Pressure flow is best visualized from still another point of view—by ignoring screw rotation. The screw channel is a continuous helical conduit through which a fluid may flow in the presence of a pressure gradient, normally encountered in extrusion. In analogy to die flow, product adheres to the extruder surfaces, where the velocity is zero, and flows by yielding to a maximum velocity in the middle of the channel, with a velocity profile like that in Figure 8. Pressure flow is driven by the pressure gradient and resisted by product viscosity, resulting in an expression identical to that of die flow:

$$Q_p = -\frac{\beta}{\mu} \frac{dP}{dl},\tag{7}$$

where Q_p = volumetric pressure flow, dP/dl = pressure gradient along barrel length, and β = pressure flow geometry factor.

The geometry factor is analogous to the die constant, and the negative sign indicates that the flow direction is opposite to that of the pressure gradient (product moves toward lower pressure). As in the die constants, the geometry factor is affected strongly by channel cross-section dimensions and less so by channel length. For shallow rectangular flights, it is:

$$\beta = \frac{1}{12} \pi D h^3 (1 - \frac{ne}{t}) \sin^2\phi .$$ (8)

More rigorous equations are available in the references listed for the drag flow equation, with the same limitations.

In the absence of significant flight-tip leakage flow, the net extruder flow is simply the sum of drag and pressure flows:

$$Q_{net} = Q_d + Q_p .$$ (9)

The velocity profiles of drag and pressure flow may also be added or superimposed to obtain the net velocity distribution shown in Figure 8. In the common case of pressure buildup along the screw, pressure flow is negative, so there is a zone of backward flow near the screw root (in closed discharge, pressure flow equals drag flow, and the forward and backward flow zones are equal), with an intermediate stagnant point. This singularity can lead to problems, especially in heat transfer, where the stagnation point can be a focus of scorching. In cases where the pressure diminishes along the screw, the pressure flow becomes positive and augments the drag flow, with no negative flow zone.

Flight-Tip Leakage

The third flow component in single-screw extrusion is flight-tip leakage, which can be significant with high pressure gradients, low viscosities, and/or loose-fitting or grooved barrels. This is caused partially by drag, analogous to cross-channel drag flow, but mainly by pressure drop across the flight tips. As in the case of drag and pressure flows, equations in the literature are available to calculate leakage flow but are not particularly useful for food extrusion, due to deviations from ideal fluid behavior and to complex mechanical designs. The leakage can be very significant, however, particularly with respect to heat generation. Since leakage can substantially reduce overall flow, any heat generated in the extruder is applied to a smaller mass, resulting in greater temperature rise. Furthermore, material flowing over the flight tips is subjected to a very high shear rate, due to a combination of high tip velocity and small tip clearance creating high intensity conditions of mixing and heat generation. Flow between flights by leakage can also disrupt the flow circulation pattern within the screw channel, mitigating the stagnation effects and providing better overall mixing—but creating a wider residence time distribution. In cooking extrusion, these effects can be positive and are often enhanced by design. For forming, however, leakage is generally to be minimized, especially when product cooling is required.

COROTATING TWIN-SCREW EXTRUSION

Because the corotating twin is the most complex of the three styles, both mechanically and in flow mechanism, we turn to it last. As shown

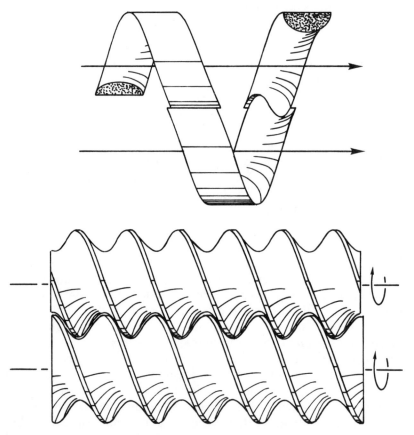

Figure 10. Corotating twin-screw extruder. Bottom, intermeshing screws; top, channel around both screws formed by screw flights and barrel surface. Note the change in the channel shape as it moves from one screw to the other.

in Figure 10, it is superficially similar to the counterrotating twin in that the screws intermesh and interact. It is different, however, in that the screws are pitched and rotate in the same direction, although still fully intermeshing in that all screw surfaces are in contact with the adjacent screw or barrel. The most obvious difference is that the corotating style has a continuous channel running the length of the screw, similar to that of the single-screw extruder. The discrete chambers of counterrotating models are not found here. A difference from the single, however, is that the continuous channel is wrapped around *both* screws and changes shape at each transition from one screw to the other. This transition causes a redistribution of the product and also acts as a bottleneck, retarding pressure flow. Thus, the screw is indeed divided into segments, but they are in communication via a continuous channel running through them.

The flow behavior of the corotating twin is therefore a combination of the other two styles—the extruder acts as a shear pump with some positive displacement characteristics. Either of these tendencies can be enhanced by screw design for any particular application, making the corotating extruder very flexible. These variations are discussed further in the following sections.

Mixing, Pressure, and Heat Generation

We have already explored the flow circulation pattern in the single-screw extruder. For this pattern to develop, the screw must be full, with all surfaces wetted by the product. However, many extruders operate with a starved-feed section, in which an open-air channel exists in front of the product, preventing pressure generation. Except in open-discharge situations, pressure is required to force the product through a die. In single-screw extrusion, this pressure is transmitted back through the screw channel to the choke point, where the screw first becomes full. This is often accomplished and controlled by the use of compression screws, where the pitch and/or channel depth are reduced along the screw. Under constant conditions, the pressure then rises steadily along the screw, and the normal circulation pattern develops. With further screw compression, however, the pressure often reaches a peak substantially before the die, after which it declines. This is because the drag-flow component declines with compression. With the total flow rate fixed by the previous screw geometry, pressure flow then becomes positive, augmenting drag flow, as indicated in the lower right velocity profile in Figure 8. This configuration is useful where the desired cooking pressure is greater than the desired die pressure. It is also useful for heat transfer, where high product velocity near the barrel surface, reduction of product bed thickness, and reduction of product stagnation are all important.

Pressure and Velocity Distribution

As shown in Figure 11, ciculation within the counterrotating twin-screw extruder chambers is somewhat different from that in the single

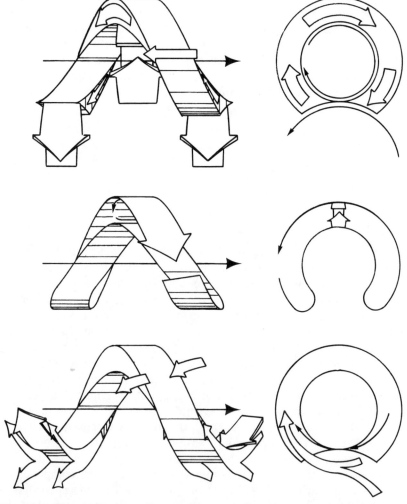

Figure 11. Flow components in the counterrotating twin-screw extruder. Top, mechanical forces acting on surfaces of material in the screw chamber; center, recirculation patterns (mixing) in the screw chamber caused by mechanical forces; bottom, leakage flows into and out of the screw chamber.

screw. The chambers are filled with product in contact with moving extruder surfaces, which shear the product and cause it to flow. Product is dragged backward by the rotating screw (as in the single-screw case) until it impinges on the interfering flight from the adjacent screw, where it is turned outward into a returning stream. Cross-flow circulation is also generated by the relative barrel motion.

To visualize the velocity distribution better, it is helpful to take the point of view used for single-screw extrusion, i.e., a stationary screw and rotating barrel. This is more difficult to picture in the twin-screw case, but can be accomplished by picturing one of the screws in the center of a planetary system—the second screw rolls around the fixed unit along with the barrel. If we now look at a portion of the chamber some distance from the interposed flight tip, we see the same geometry as in the single-screw channel, and find that the same drag-flow and pressure-flow conditions exist. Along with normal drag flow, however, an additional push is provided by the interposed flight tip, which travels in the downstream direction. The push is transmitted to the product as a negative pressure gradient, since pressure is highest at the upstream end of the chamber. This creates a positive pressure flow and a velocity profile like that at the bottom right of Figure 8. As noted for the single-screw case, such a profile is especially good for heat transfer.

With a downstream pressure drop in each chamber, we find that the pressure does not rise steadily along the extruder, as in the single-screw case, but in a zig-zag fashion. The average pressure does rise steadily as each chamber progresses along the screw, however.

Pressure is not necessarily transmitted all the way back to the beginning of the screw, as it is under steady conditions in the single-screw extruder. In the ideal case, rate is fixed by volumetric capacity and is independent of pressure—only one chamber is needed to generate any pressure. However, depending on leakage, several chambers in series are required to generate a particular pressure. At higher operating pressure, the pressure merely begins to rise earlier in the extruder, with little effect on rate. Upstream of this point, the chambers are not completely filled and do not contribute to the pressure rise.

Leakage Flow

Leakage occurs over the flight tips, as in single-screw extrusion, and also through the screw gaps by both pressure and drag. The pressure gradient is generally much higher across the flights than for the single-screw, due to the capacity for higher overall pressure gradients and because of the pressure drop in each chamber (the pressure "zig" in one chamber is opposite the "zag" in the next). The gradient is minimized, however, by closer tolerances and the absence of grooves. A more important leak is found in the "calender gap," where the screws roll on each other. Here, a substantial drag is added to the pressure gradient as both screws pull product through the gap, creating very high point pressures. This limits some applications of the counter-rotating twin. To some extent such leaks are beneficial, however, as they provide the only route for mixing product between screws, which is helpful for product uniformity.

Overall Flow Pattern

As previously noted, the corotating twin-screw extruder combines features of the other two styles and manifests elements of both in its flow pattern, shown in Figure 12. As in the case of the counterrotating twin, flight tips from the adjacent screw are interposed in each screw channel, creating a series of chambers in which product circulates. This circulation is not complete because the chambers are not isolated— a part of the product can transfer between chambers instead of recirculating—but is assisted by the motion of the interposed tip, which moves in the direction of recirculation instead of pulling the product out of the chamber, as in the counterrotating calender gap. Cross-channel circulation is also generated in the chamber.

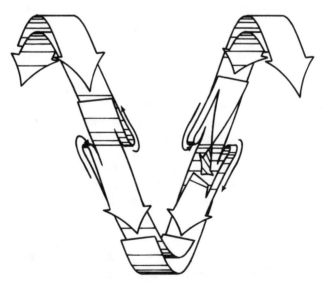

Figure 12. Flow components in the corotating twin-screw extruder. Recirculation in each chamber is augmented by leakage flow between chambers, with product turnover at transfer points.

Since the interposed flight tips do not completely seal off the chambers from one another, a substantial flow moves between them. As in the single-screw case, pressure and drag flow components can set up positive and negative flow zones throughout the channel as the product moves around the periphery of both screws. The outer flow path is generally smooth and positive, resembling the single-screw drag flow (with some assistance from the interposed flight tip as in the counterrotating case). The inner flow path reveals an important feature of the corotating twin, however, in that it does not transfer smoothly. The shape of the channel changes at the screw intersection, causing both a complete disruption of the normal flow pattern and mixing of the product. In addition to the shape change, the close-fitting interposed flight scrapes up product from all screw surfaces to redistribute stagnant product into the main stream flow. Thus both midstream and screw-surface stagnation are eliminated, permitting more uniform processing and high heat transfer rates with less danger of scorching.

Machine Geometry

The relative magnitudes of the flow components (circulation, forward and backward flow) depend, of course, on pressure gradient and product viscosity but are also strongly affected by machine geometry. Figures 10 and 12 illustrate a two-lead design, commonly used for cereal processing. This design permits substantial transfer flow—the interposed tips do not dam up much of the screw channel. In single-lead designs, the flight tips are generally wider, so the process takes on more of a positive-displacement character and can generate higher pressure gradients with less product transfer between screws. Triple-lead designs are also available. Flight depth also affects behavior. Deep flights provide good extensive mixing and high volumetric capacity, at the expense of shear (intensive mixing) and pressure generation, as they permit easier product interchange between screws.

OPERATING LINE VS EXTRUDER STYLE

Returning briefly to the operating line concept, we may now see how overall process behavior may be reflected in the various extruder styles. The counterrotating twin-screw extruder, being insensitive to pressure changes, exhibits an essentially flat operating line—flow rate is constant at any pressure. The situation is very different with single-screw extrusion, where pressure flow is significant. For open discharge, the overall flow is equal to the drag flow, which does not vary with pressure or product characteristics but is a fixed function of geometry and screw speed. As pressure increases, however, so does the negative pressure-flow component, eventually becoming equal to, and cancelling out, the drag flow. The elasticity or slope of the flow rate then depends on the pressure-flow component, which is a function of viscosity and geometry.

Comparing Equations 5 and 7, we see that the ratio of pressure flow to drag flow increases with the relative channel depth (h/D) and pitch and decreases with screw speed and viscosity. These factors similarly increase the slope of the operating line—screws with shallow channels, short pitch, and high speed are less sensitive to pressure changes, especially when processing high-viscosity materials. Ironically, depending on the operating region, viscosity can have either a positive or negative effect on flow rate. The screw functions more efficiently with higher viscosities, which create more resistance at the die. Corotating twin-screw extruders, having an intermediate flow mechanism, are generally not as sensitive as the single-screw extruder to pressure change and usually have flatter operating lines, depending on the particular screws used.

RESIDENCE TIME

In Chapter 3, the concept of residence time distribution was introduced for comparison of continuous processes. Having explored the three extruder flow mechanisms, we may now explain their influence on residence time. The perfectly sealed counterrotating twin-screw extruder comes close to the ideal plug-flow condition; with no leakage, the total range of residence time is that required to empty one chamber, usually only a small fraction of the total time elapsed in the extruder. Leakage tends to spread the range out by intermixing product from different locations along the screw.

The single-screw extruder is, of course, subject to high leakage rates with large pressure gradients—in the extreme case of nearly closed discharge, the single-screw extruder approaches the 100% backmix condition. More important, however, even with open discharge (no pressure flow), the drag-flow mechanism leaves a stagnant layer of material on the screw surfaces so that the residence time distribution is inherently broad.

By scraping up the stagnant product layer, the corotating twin reduces its effect on residence time distribution, making the distribution generally narrower. However, with very high pressure gradients, often created by the use of mixing elements and reverse-pitch screws, back flow becomes the dominant factor in residence time distribution, so the distinction of self-wiping is lost. Furthermore, back flow is a common occurrence even in the absence of an imposed pressure because of the chamber-generated gradient, depending on geometry. In general, therefore, the residence time distribution of the corotating twin can vary widely between those of the counterrotating twin-screw and the single-screw extruders, depending on process conditions and machine design.

HEAT GENERATION

Heat is generated in all extrusion processes from shearing of the viscous product stream. As outlined later, this is put to good use in cooking processes but is usually minimized in forming operations, which often require product cooling. The amount of heat generated is reflected by drive motor power, which is converted to thermal energy by viscous dissipation of the intermediate mechanical energy of the moving machine parts. Total energy transfer is a product of screw speed, shear stress (viscosity and screw speed), and surface area (extruder diameter and length), which may be described approximately by:

$$Z = \pi^3 D^2 N^2 \mu \left(\frac{\pi D}{2h} + \frac{ne}{\delta} \right) l, \qquad (10)$$

where Z = extruder drive power, δ = clearance between barrel and flight tip, and l = wetted screw length (after choke point).

This equation describes the total energy input. Of more interest to the food technologist is the specific energy, usually reported as kilowatts per kilogram, which is the amount of energy absorbed per unit of product and is found by dividing the energy flow by the product mass flow.

By its nature as a shear pump, the single-screw extruder operates with high specific energy. Except in starved-flow sections, all surfaces are wetted as pressure is transmitted back from the die throughout the screw length. In high-pressure applications, the overall product flow is drastically reduced relative to power consumption. (In closed discharge, specific energy becomes infinite.) This effect is even greater when leakage flows are enhanced by barrel grooves.

For similar screw size and speed, the power consumption of the counterrotating twin is about the same as that of the single over the same wetted screw length. Its throughput, however, is larger, greatly reducing specific energy. Indeed, the throughput is about double that of the single-screw in open discharge, halving the specific energy. This difference is even more pronounced at high pressure, which does not strongly affect counterrotating throughput. Furthermore, its wetted screw length is limited—heat is not generated in any significant quantity until the pressure generation point is reached. For these reasons, counterrotating twin-screw extruders are not useful when high specific energy is required but can be very productive for heat-exchange processes.

Corotating twin-screw extruders again fall between these extremes. In open discharge, very low specific energy is the norm. At high pressure, however, specific energy is increased by retarding the flow rate and increasing the wetted length, although to a lesser extent than in the single-screw case. These functions may also be performed by mixing elements, reverse-pitch screws (Tayeb et al, 1988), and/or restrictions within the barrel, which create zones of high energy absorption at desired locations along the screw and make the unit very flexible over a range of process conditions.

OPERATING LINE VS ENERGY ABSORPTION

We noted earlier that the slope of the operating line is greater at lower viscosities. When the specific energy input is permitted to affect product temperature (in the absence of thermal control), we expect to see changes in viscosity as well. As we move toward higher pressure, more of the input energy is converted to temperature rise, which

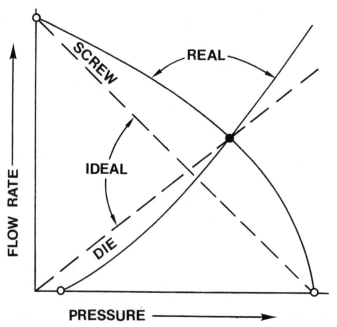

Figure 13. Deviation from ideal behavior in extruder operating lines.

normally reduces product viscosity and gives a corresponding increase in the operating line slope. In real situations, therefore, we expect the line to depart from the linear ideal with a downward curve, as shown in Figure 13. Actual rheological properties of cereal products exhibit even more complex behavior, leading to other variations in the operating lines that are best determined experimentally for any particular extruder and product.

A further complication in the prediction of extruder operating lines is encountered in the common practice of starved feeding. Here, the wetted screw length is not fixed but varies with pressure by an upstream movement of the choke point. Hence, the rate may be independent of pressure over a portion of the operating range—instead of a rate decrease, the power consumption (and temperature) rise as more product is sheared by the machine surfaces.

One additional aspect of extruder behavior should be introduced before leaving the operating line construction: stability. When the system operating point is near the closed discharge end of the operating line, the overall flow rate becomes very sensitive to small changes in pressure, especially when the slope is large, as in the temperature-sensitive viscosity case. This can lead to surging, a problem that is common in single-screw extrusion but seldom found in the twin-screw styles since they do not normally operate near this danger zone.

Extrusion-Forming of Cereal Products

Forming extruders transform a cooked, moist (20–30% moisture) feed material into a dense, relatively homogeneous product of desired shape for further processing. The shaping may be accomplished by the extruder itself by die-face cutting, or other methods may be used to augment the product shape after it emerges from the die. Extruded pellets may be transformed by flaking or shredding, or continuous extruded product streams may be manipulated by additional machinery (embossing, folding, laminating, cutting) to create more complex shapes.

For each of these applications, optimum physical properties (adhesiveness, flexibility, plasticity, elasticity) are created by the extruder, through formulation and cooking conditions. The extruder also compresses, deaerates, and homogenizes the product to improve appearance and uniformity.

EXTRUSION EQUIPMENT AND PROCESSES

Extruders used for forming processes normally operate under low-shear conditions, resulting from deep flights and/or low screw speeds, to minimize product heating and starch degradation. These goals are also met by cooling the product by conduction via barrel jackets or cored screws, usually with cool tap water but sometimes with chilled heat-transfer fluids. Cooling also makes the product easier to handle after extrusion, by increasing its viscosity so that it holds its shape better and by reducing stickiness. With high in-process viscosities and usually very restrictive dies, forming extruders operate at high pressure, often exceeding 700 N/cm^2 (1,000 psi).

The original forming-extrusion processes borrowed technology from the plastics and ceramics industries, finding use first in pasta operations and then for breakfast cereals. Single-screw extruders, used widely for this purpose today, have changed little from these early machines. They consist of relatively short screws (L/D about 10) rotating in close-fitting grooved barrels, with water-cooling jackets and sometimes cooled screws and dies. A modest compression of about 3:1 is often used, to exclude air from the product and to improve heat-transfer efficiency. To avoid generating excess heat, compression and barrel grooving must not be carried too far.

The feed to a forming extruder is the product of any of the cooking processes outlined in Chapter 3. Since forming is done continuously, a continuous cooking process—particularly extrusion cooking—is preferred for process stability and product uniformity. When it follows cooking extrusion, the forming unit may be an integral part of the cooker, or it may stand alone. In the first case, both operations are done in separate zones in one machine. Many prefer to use the older method requiring separate machines, however, because it is easier to optimize each process separately for the best overall productivity and product quality. This is done at the expense of a more complex operation, with more attention needed for process control.

Preferred Extruder Styles

Free-standing forming extruders usually use the simple single-screw design, choke-fed and run at the minimum screw speed required to attain the desired rate and pressure without overworking the product. Overwork and air incorporation must be avoided in the feed hopper also. The hopper often employs rotating paddles to convey the hot, sticky feed material into the screw flights. Feeding efficiency is generally improved by cooling the barrel in the feed zone. Due to the poor mixing action of the single-screw extruder, cooling must be done carefully to avoid heterogeneous processing or hot spots in the product, a situation exacerbated by the continuous heat generation of a shear-intensive process.

Free-standing single-screw forming extruders are available from APV Baker, Bonnot, and Wenger and as ancillary units for continuous cookers from Readco and Textruder. Plastics machines such as those from Egan and Farrel may also be adapted and often may be found in used condition at low cost.

Corotating twin-screw extruders are more efficient at both heat transfer and pumping, with relatively little heat generation. Due to their high price, however, they are seldom used as free-standing forming extruders. In most cases, the cheaper single-screw machine is sufficient. Corotating extruders may be used for forming if desired, and a counter-rotating twin-screw former is available from Wenger. In integrated processes where cooking and forming are done in the same extruder, the cost difference is mitigated by the additional convenience and efficiency of the twin, so that more of these units have come on-stream in recent years.

To fit both processes into a single extruder, very long screw lengths are required (L/D up to 30). Most of the length is taken up for cooking, followed by a shorter, cooled forming section separated by a vent. Venting is important for rapid removal of cooking heat before the product enters the forming zone. When the product flashes off its excess heat in the form of steam, it can cool from a cooking temperature above 180°C to near the boiling point (100°C) or lower, if vacuum is applied to the vent. In addition to allowing room for the two steps, long screws also permit better separation of the operations, with less heat transfer between them. The twin-screw mechanism is particularly good at separating process zones by mechanical design, making it a good choice for integrated operations. It also permits starved running of the forming section without overworking the product. This is a necessary condition since the rates in the two sections must match— the forming screw speed cannot be independently controlled and must exceed the minimum for the production rate. In this case, the extruder capacity is controlled by wetted screw length, which floats in equilibrium with the feed rate and other variables such as viscosity. Twin-screw systems are available from APV Baker, Buhler Inc., Clextral, Wenger, and Werner & Pfleiderer, among others.

Ancillary Processes

The extruder itself is only the first step in the forming operation. Its job is to deliver a cooked product of the desired consistency and temperature to the point at which the shape is actually created and to generate sufficient pressure to maintain the desired production rate. A wide choice of additional steps is available to create shapes by further manipulation the extruded product. Selection of the best arrangement for any particular product depends, of course, on desired appearance, tempered by practical considerations. These include the propensity for product breakage (peninsulas and isthmuses tend to become archipelagos), density (hollow shapes take up more room), and recognizability (one should avoid the temptation to become an extruder sculptor, as simple shapes are easier for the consumer to relate to).

EXTRUSION DIES

Extrusion dies constitute the first step in product shaping. Pieces may be cut off at the die face into a final shape before puffing or into a pellet for flaking or shredding, or product may be drawn or pulled off in a continuous stream for further shaping.

The cross section within the die is a two-dimensional shape, depending on the die design. However, as the product emerges from the die, distortions caused by the flow velocity profile disrupt its neat, constrained form. Distortion must be designed for and may be enhanced or diminished as desired for the final product. In cases where the distortion is to be minimized, elastic effects may be reduced by using long, tapered inlets to the final die opening. In extreme cases (for maximum definition of a complex shape), the velocity profile may be flattened by using smooth die surfaces to cause slipping.

For die-face-cut products, however, distortion may produce a more attractive product, due to the less "artificial" look of a rounded piece. Rounding may be accentuated by using short dies with minimum taper and with rough surfaces (normal tool marks) to eliminate slip and develop a full velocity profile. Wall temperature is another important factor in die shaping. Cooling (typically with tap water) increases product viscosity at the die surfaces, for an increase in the steepness of the velocity profile. This method has the advantage of providing continuous control during processing and the flexibility of allowing differential control between portions of the same die for special effects.

Differential flow in a die may also be controlled by varying the land length. Longer flow paths create more resistance and slow the velocity where used. This is most useful in creating three-dimensional effects in die-face-cut products or in eliminating too-thick or too-thin sections in complex shapes.

Continuous extrusion for postextrusion forming normally requires a uniform velocity distribution to avoid stretching or tearing of the product as it is pulled from the extruder. In the simple case of a slit die for continuous ribbon-forming, additional drag is created by the ends of the slit, reducing the velocity at these points. To assure uniform product thickness after it is pulled from the extruder, this end-effect may be eliminated by enlarging the thickness of the slit at its extremities or, preferably, by machining an additional taper at the slit ends so that the land length is reduced to about half that of the main portion of the slit.

Later processing (e.g., toasting) further distorts product shape due to thermal stress from uneven heat transfer to various product surfaces. For this reason, die design must be derived from the desired shape of the final product after correcting for three levels of distortion— die-flow characteristics, postextrusion forming, and thermal warping. Except in the simplest of cases, this is a trial and error exercise, usually taking at least three attempts. The best strategy is to make a "guesstimate" of the die shape based on experience with similar products for the first trial. The procedures shown in the accompanying box usually converge quickly on the final design.

Standard dies are available from extruder manufacturers and from specialists in the field, but usually new die development is left to the user.

DIE-FACE CUTTING

Being a continuous process, extrusion creates a steady stream of product that must be subdivided into pieces of suitable size for breakfast cereal. This may be done some time after extrusion (see "Postextrusion Forming") or as the product emerges from the extruder by die-face cutting. Die-face cutting uses a rapidly moving knife blade passing in close proximity to the die orifice. The blade clips off small segments of the continuous product stream and converts it into a shower of pieces—confined by a metal chamber (frequently perforated for ventilation) that also directs the product onto a collecting conveyor.

Cutting blades are mounted in rotating supports direct-driven by variable-speed motors at speeds ranging from several hundred to several thousand revolutions per minute (rpm). The overall layout of the cutting system varies among extruder manufacturers and product applications, as shown in Figure 14. Coaxial cutters provide the most uniform cutting motion for multiple-orifice dies laid out in a circular pattern about the extruder axis. The motor must be isolated from the product stream by a drive shaft, to protect both product and motor. This may be cantilevered by outboard bearings, resulting in a clean design with no mechanical supports to clutter up the product flow. In using this design, however, die-knife clearance is somewhat limited by the possibility of shaft whipping. It also requires extra heavy cutter support to assure rigidity at the bearings. By mounting the cutter directly on the die plate via an integral shaft and bearing, these problems may be eliminated—at the risk of flow interference and product contamination from the additional mechanism and from necessary lubrication within the collecting chamber. With both variations, blade velocity is limited by the small radius of the cutter.

With outboard cutter-drive mounting, the motor may be easily isolated from the product. In parallel shaft arrangements, the rotating cutter swings through a large radius, attaining greater knife velocity than a coaxial cutter at the same rpm. A mixed blessing, the higher velocity increases the chances of product damage within the cutting chamber if the cut pieces impinge on the whirling blades as they cascade to the collecting conveyor—a probability increased by the necessarily larger chamber and increased turbulence. Long outboard cutter arms

Iterative Development of Extruder Die Design

1. Initial shape "guesstimate"

a. Using circular dies, experimentally determine overall die resistance (number and size of dies) that provides the best product/process behavior (flow rate, texture, stickiness). Calculate the total die resistance using the equations in Figure 4 and the expansion ratio (ratio of product and die cross-section areas).

b. Prepare a scale drawing of the cross section of the desired final product shape.

c. Prepare a second cross-sectional drawing with corrections for distortion. This is accomplished by reducing the area of the cross section from that of the desired product by the expansion ratio. The reduction is not done uniformly except in simple cases such as for radially symmetrical shapes. More reduction in thicker zones is called for, as in the contrast between an inflated and a deflated balloon.

d. Calculate a die constant for the new design either by using the general equation or by dividing the shape into simple segments where the more exact equations may be applied separately and then added together.

e. Fabricate the new die and assemble it in the extruder along with other dies of similar die constant. All dies should have about the same resistance—the new die may be used along with a number of roughly equivalent circular dies. This is so that the total resistance matches that found in step a.

f. Test by running through the entire process to operate all distortion factors. Data should include flow rate from the test and circular dies, as well as the product shape and its quality and process behavior. This might be repeated several times with different circular dies to obtain the best process conditions.

2. Extrapolation from initial results

a. Compare the actual and desired product for size and shape. Determine portions of the product cross section and thickness that need to be altered.

also limit die clearance by their tendency to wobble. While satisfactory for single-orifice extrusion, the outboard drive arrangement may not be suitable for multiple orifices since the cutting motion varies among orifices at different locations.

In selecting a blade arrangement for any particular application, the first consideration is, of course, assuring that the cutting frequency produces pieces of the correct size. The cutter speed needed for this is a function of flow rate and number of cutting blades and die orifices:

$$N_c = \frac{G}{n_d \, n_b \, g} \,, \tag{11}$$

where N_c = cutter speed (rpm), G = total flow rate (mass/min), n_d =

---

b. Prepare a new die drawing with dimensions altered from the original die by the ratio of desired to actual product dimensions. It is best to overshoot on these alterations to bracket the final dimensions, which then can be interpolated more accurately.

c. Alter the lead-in portion of the die to change relative thickness where desired. Product thickness may be increased by chamfering the die to decrease land length, resulting in greater velocity.

d. Determine the actual die constant of the original die by comparing its flow rate to that of the circular dies:

$$k_e = k_c \, (Q_e/Q_c) \,,$$

where subscripts e and c = experimental and circular dies, respectively, assuming that the pressure and viscosity are the same in both cases. This is a good assumption if the die flows are about the same.

e. Estimate the die constant for the altered die by applying the equations and correcting for deviations from this calculated value by the ratio of actual to calculated constants for the initial design.

f. Repeat the tests outlined in steps 1e and 1f to generate experimental data with the altered design.

3. Final die design

In many cases, two trials are sufficient to evolve a new shape. More frequently, however, a third trial is needed to solve a complex new shaping problem, and such a trial should be planned. The procedure outlined in step 2 is followed except that, with luck, the alterations will be smaller and the new dimensions will fall between those already tested, so that they may be accurately predicted.

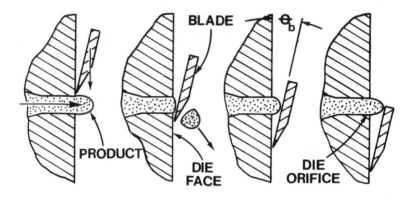

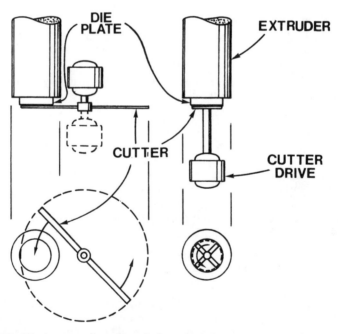

Figure 14. Die-face cutting. Top (left to right), cutter sequence at die face. Θ_b = blade angle. Bottom, various cutter layouts: left, parallel axis arrangement (cutter drive may be in either inboard or outboard position); right, coaxial position.

number of die orifices, n_b = number of cutter blades, and g = individual-piece weight.

From this, it is apparent that the cutter can operate over a wide range of speeds to obtain the correct piece weight, depending on the number of blades used. In practice, other factors must be considered to find the best speed.

To obtain a good, sharp cutoff of the product, a reasonably high blade velocity must be used. At high velocities, the product shears more like a solid due to its inertia with respect to the moving blade; at low velocities, contact time with the product is longer, allowing it to distort and, in the extreme, to stick to the blade. Shearing of

Figure 15. Coaxial die-face cutter with curved blade. Holes in the die plate are for die inserts containing the actual die orifices. (Courtesy Anderson International Corp., Cleveland, OH)

the product between the moving blade and stationary die is especially important in forming easily distorted hollow shapes and is also a function of clearance—closer blades cut more cleanly. To further improve shearing, the blades may be angled somewhat from the radius of rotation, and sometimes are curved (Figure 15) so that this angle increases with radius—this introduces a slicing component to the blade motion.

Although high velocity provides cleaner cutting, the cutter can damage the pieces after they are cut by catapulting them at high speed against other machine surfaces. The best speed for a particular application must take this into account.

Another factor in choosing blade speed is the desired squareness of the cut. At low blade velocities, the cut is oblique due to the relatively greater product movement from beginning to end of the cut. The oblique angle may be calculated by equation 14, derived later for blade clearance.

Blade velocity depends on cutter diameter as well as speed:

$$v_b = \pi D_c N_c ,\tag{12}$$

where v_b = blade velocity and D_c = cutter diameter. Coaxial cutters generally have relatively small diameters, and can operate at high speed without generating excess velocity.

It is generally desirable for cutting blades to be as thin and narrow as possible to minimize interaction with the product for clean cutting. The blades must be strong enough to withstand stresses and must contain enough material to hold an edge for a reasonable time. Another way of minimizing blade drag is to fabricate blades so that their sides are not parallel to the die face, but angle away at the trailing edge.

Although the cutters remove discrete segments of the emerging product stream, the stream is continuous and begins to protrude from the die face as soon as the blade edge passes. To reduce product drag and minimize distortion, a space for this protrusion must be provided in the trailing region of the blade by angling it away from the die face. The minimum clearance angle depends on the relative blade and product velocities. Since the product is assumed to behave as a fluid in the die, it has a maximum velocity, which in the case of ideal fluids in circular dies is twice the average velocity. (These assumptions are conservative in that other geometries, slip, and the pseudoplasticity normally exhibited by cereals tend to reduce this maximum.) Average velocity v in the die is:

$$\bar{v} = \frac{G}{n_d \rho A_d} ,\tag{13}$$

where ρ = product density (usually about 1.2–1.3 g/cm^2) and A_d = die orifice cross-section area.

Combining this with Equation 12, and correcting for the velocity maximum, the minimum blade clearance angle becomes:

$$\theta_b = \tan^{-1} \frac{2 g n_b}{\pi \rho A_d D_c} \tag{14}$$

where θ_b = blade angle. This angle should not be exceeded by too much, since the outer blade surface also contacts the product: a very steep angle would require that the blade be quite thick or else angled on the outer surface. Either of these alternatives transmits excess velocity to the cut pieces.

POSTEXTRUSION FORMING

Die-face cutting is an efficient method of creating cereal product shapes, but it is generally limited to simple two-dimensional products, with variations caused by flow distortion. For more complex shapes, postextrusion forming is the rule. Shredding and flaking, which may be considered processes of this type, are discussed in other sections of this book. However, extending the extrusion process by the use of other equipment designed to modify the shape of the extruded product is another alternative.

The product of a forming extruder is generally soft and amenable to shaping by application of force, if done soon after it emerges. It is usually taken from the extruder as a continuous ribbon under tension provided by rollers or conveyors. It can then be embossed with a pattern such as a waffle-grid, bent, stuck to other streams by pressure, and subdivided into cereal-size pieces (e.g., McKown, et al, 1969). Machinery for postextrusion forming is usually specific to a particular product, and developed along with the associated extrusion dies. The process is mechanically more straightforward than die-forming because it is not affected by complex flow distortions. However, distortions do occur in further processing such as toasting and must be considered. This is especially evident in the curling that is caused by uneven response to heat treatment. Curling is often desired and can be increased by creating a differential in product form (e.g., by embossing only one side of a flat piece).

When more than one product layer is created, as in shredded wheat and postextrusion laminated ribbons, differential heat response can create a pillow shape as the exterior surfaces expand more in the higher-temperature microenvironment of the toasting process. Lamination poses more stringent requirements on the extruded product—it must be soft and adhesive enough to stick together but must not adhere to forming-machine surfaces. The same problem arises with any postextrusion handling and is eased by the use of modern nonstick coatings. Formulation and cooking conditions must also be adjusted to arrive at the best degree of adhesiveness and viscosity. The time and ambient conditions to which the extruded product is exposed in transit to the forming machine also affect consistency—the product becomes stiffer and less adhesive as it cools and dries.

SPECIAL PROCESSES

Other means of creating unique product configurations with extrusion are under continuous development. These include coextrusion, which laminates two or more product streams within the extrusion die. This has the advantage of eliminating complicated and sometimes troublesome postextrusion forming equipment. At the expense of a more complicated extrusion process, the two streams may be welded more firmly together, and in more complex configurations, depending on the imagination of the die designer. In practice, they are brought together at the entrance of the die, after which they travel together in laminar flow through the restrictive die passages (Rossen and Miller, 1974).

In both coextrusion and postextrusion lamination, two different formulations are used. These may differ only in color for visual contrast, or they may be made of different basic ingredients for taste and texture contrast. If the formulations are functionally different, interesting differential behavior in further processing can be developed.

PROCESS DYNAMICS AND CONTROL

Forming extrusion depends greatly on the feed material for much of its control. In most cases, the feed consistency is controlled by the cooking operation and the ambient atmosphere to which the product is exposed enroute to the forming extruder. Further ingredients are sometimes added between the cooking and forming steps, including flavors and nutritional fortification, which could be damaged in the cooking step.

Other than changes in the basic formulation, the variables remaining in forming extrusion to correct for changes in the feed stream during operation are screw speed and cooling rate. Screw speed in the free-standing forming extruder is usually dedicated to matching production rate to the other processes while maintaining the correct level of feed material in the feed hopper (slightly starved or choked without overflowing), so that it is not available to control other factors. In the integrated process, it is even less available, since the screw speed is selected for optimum cooking in the upstream portion of the extruder. Cooling, therefore, is the main control element.

Cooling can be applied incrementally by four means: barrel jackets in the feed section or in later sections, cored screw, or die jackets. Operating control can be achieved by varying the coolant flow rate or temperature in any of these areas, with different results from each. Cooling with feed section barrel jackets is the normal choice for quickly reducing product temperature to arrest cooking reactions and to decrease product shear by improving pumping efficiency. Where the jacket is insufficient to remove heat quickly enough, as is sometimes the case in large extruders, the heat exchange area can be increased by using cored screws. However, this has the side effect of reducing pumping efficiency and must be applied carefully.

Downstream barrel cooling can also extend the heat-exchange area. This is especially effective with compression screws that create excess heat through shear but provide a shallower product bed for better heat transfer. When downstream cooling is used, it is desirable to cool the feed section as well, so that it can build up sufficient pressure to force material into the later compression without overworking the product. Die cooling is not particularly effective at reducing product temperature due to the very short residence time available. It can create enough of a temperature gradient in the product to affect the flow pattern and can cool the product surface for easier handling. However, too much downstream cooling (including at the die) can increase product temperature at upstream points by forcing the use of higher screw speed to maintain the production rate at higher pressure, causing more shear and heat generation.

SCALEUP

In taking a newly developed cereal product into production, problems arise in providing the identical process environment on a larger scale, especially in extrusion operations that perform numerous process steps simultaneously. We have already discussed die design variations and (using the outlined procedure) can predict the number and size of die orifices needed to duplicate die pressure at a higher flow rate. This is the first step in extruder scaleup, and it assumes that the product has identical consistency in both cases—the main object of the scaleup.

The next step in the scale-up procedure is to select a larger size extruder to produce the same product at a higher rate. Here we encounter the situation that all of the factors affecting the product do not scale at the same rate. From examination of the flow rate and energy equations given here and in Chapter 3, and in comparing geometrically similar extruders (where all dimensions are in the same proportions) of different sizes operating at the same rpm, we find that flow rate and power consumption both increase roughly in proportion to the cube of the screw diameter, while maintaining about the same shear rate and residence time. When we look at barrel heating or cooling, however, we find that the area available for heat exchange increases only as the square of the diameter (twice the diameter increases the area by only four instead of eight times). In extrusion processes such as forming, where heat transfer is important, the product cannot receive sufficient processing in the larger machine unless other changes are made. These are generally a compromise between competing concerns.

We have already discussed ways of increasing cooling, such as by using other surfaces to augment the barrel surface (i.e., cored screws) and by applying lower-temperature coolant at a higher rate. These are limited in their application, however, and introduce other changes in the process by altering flow patterns and causing higher temperature gradients in the product. Thus, it is necessary to modify either operating conditions or screw geometry, or both, to arrive at the best solution when large changes in scale are needed.

Operating conditions may be modified by reducing the screw speed in roughly inverse proportion to the diameter (twice the diameter operating at about half the speed). Unfortunately, other changes in the process occur at the same time: the residence time in the extruder increases in proportion to the diameter, and the shear rate and specific energy (mechanical energy per unit of product) both decrease in inverse proportion. Where these are important, additional modifications are required. The approximate scale-up factors for geometrically similar machines operating at the same product heat-transfer rate are:

$$D_2/D_1 = \sqrt{Q_2/Q_1}$$

$$N_2 = N_1 \frac{D_1}{D_2}$$

$$(q/Q)_2 = (q/Q)_1$$

$$\bar{\gamma}_2 = \bar{\gamma}_1 \frac{D_1}{D_2} \tag{15}$$

$$t_2 = t_1 \frac{D_2}{D_1}$$

$$(Z/Q)_2 = (Z/Q)_1 \frac{D_1}{D_2} ,$$

where Q_1 and Q_2 = the two different flow rates, D_1 and D_2 = the screw diameters, N_1 and N_2 = the screw speeds (rpm), $\bar{\gamma}_1$ and $\bar{\gamma}_2$ = the average shear rates, $(Z/Q)_1$ and $(Z/Q)_2$ = the specific energies, $(q/Q)_1$ and $(q/Q)_2$ = the heat transfer rates per unit of product, and t_1 and t_2 = the residence times of the two different-sized machines.

In terms of modifying screw geometry, by reducing relative screw flight depth in the larger machine, residence time may be decreased and shear rate increased while maintaining the correct heat-transfer characteristics (flow rate proportional to surface area). To achieve this, we can estimate the new channel depth and screw speed by:

$$h_2 = h_1 \left(\frac{D_2}{D_1}\right)^{2/3}$$

$$N_2 = N_1 \left(\frac{D_1}{D_2}\right)^{2/3} \text{,} \tag{16}$$

where h_1 and h_2 = the two different channel depths. These scale factors create about the same heat-transfer and specific-energy conditions at a somewhat longer residence time and reduced shear rate in the larger machine:

$$t_2 = t_1 \left(\frac{D_2}{D_1}\right)^{2/3}$$

$$\bar{\gamma}_2 = \bar{\gamma}_1 \left(\frac{D_1}{D_2}\right)^{1/3} \tag{17}$$

at a flow rate still proportional to the square of the diameter:

$$\frac{D_2}{D_1} = \sqrt{\frac{Q_2}{Q_1}} \text{.} \tag{18}$$

To maximize flow rate and use more of the available pumping capacity of the extruder, it is possible to approach proportionality to the cube of the diameter (while maintaining heat-transfer characteristics) if one can compromise further in other areas. One effective way of increasing relative surface area is to use a longer barrel—a long, thin cylinder has more surface area than a short, stubby one of equal volume. This leads to another set of approximate scale factors, for cases where the residence time and specific energy are not important:

$$\frac{D_2}{D_1} = \left(\frac{Q_2}{Q_1}\right)^{1/3}$$

$$L_2 = L_1 \left(\frac{D_2}{D_1}\right)^2$$

$$h_2 = h_1 \qquad\qquad (19)$$

$$N_2 = N_1$$

$$(q/Q)_2 = (q/Q)_1$$

$$(Z/Q)_2 = (Z/Q)_1 \frac{D_2}{D_1}$$

$$t_2 = t_1 \frac{D_2}{D_1}$$

$$\bar{\gamma}_2 = \bar{\gamma}_1 .$$

However, these factors can lead to impractical barrel lengths and are of only marginal interest because of the large increase in specific energy. This is counterproductive in a cooling process and may require excess drive power.

By using a more modest change in barrel length along with a reduction in screw speed, the specific energy may be brought into line at an intermediate production rate, leading to a third set of scale factors:

$$\frac{D_2}{D_1} = \left(\frac{Q_2}{Q_1}\right)^{0.4}$$

$$L_2 = L_1 \frac{D_2}{D_1}^{1.5}$$

$$N_2 = N_1 \sqrt{\frac{D^1}{D_2}}$$

$$(q/Q)_2 = (q/Q)_1 \qquad\qquad (20)$$

$$(Z/Q)_2 = (Z/Q)_1$$

$$t_2 = t_1 \frac{D_2}{D_1}$$

$$\bar{\gamma}_2 = \bar{\gamma}_1 \sqrt{\frac{D_1}{D_2}}$$

$$h_2 = h_1 \frac{D_2}{D_1} .$$

In choosing the correct set of scale factors for a particular product, a judgment based on the relative importance of the various factors on product quality must be made after reviewing experimental data from the pilot plant extruder. Since the latter are only approximate, they must be applied carefully, especially when a large scaleup is anticipated. Other geometric factors such as screw shape and pitch also enter into the picture. Their effects depend greatly on the rheological properties of the food and cannot easily be generalized.

This scale-up discussion based on single-screw extrusion may be applied to twin-screw extrusion as well, with some special notes. First, the twin is generally easier to scale up over a wider size range. Its flow mechanism is simpler and behaves more like the ideal assumptions made in developing the scale-up factors. Second, the second set of scale-up factors (equations 16–18), perhaps the most useful for single-screw extrusion, do not apply to the twin because its channel depth cannot easily be varied. Most twin-screw extruders are manufactured with geometrically similar screws over a wide range of sizes. Therefore, different tactics involving screw profile must be employed. Specific energy must be brought into agreement by adjusting internal resistance with screw pitch. This is a trial and error process on the larger machine, starting perhaps from a design based on one of the other sets of scale-up factors used to meet other criteria.

Direct Expansion of Cereal Products by Extrusion

The advantages of direct expansion are obvious—a virtually complete product can be made in one piece of equipment that comprises the mixing, cooking, forming, and texturizing steps and replaces a complex process line. Only light toasting and optional coating are needed after extrusion. The extruder also acts as a dryer, reducing the need for toasting.

In principle, extrusion puffing creates a gelatinized product under pressure with sufficient moisture and temperature so that the moisture flashes into steam within the product when the pressure is released. The pressure release must be done quickly, and the product must be homogeneous and elastic so that it inflates into a relatively uniform foam structure. After flashing, the product quickly "sets" in its expanded form due to both moisture loss and the associated rapid evaporative cooling. Feed moisture to the extruder is generally in the 15–25% range, which drops off by up to 8% upon flashing. Extrusion temperature commonly exceeds 180°C, depending on formulation and the desired degree of expansion.

To reach these high temperatures, conductive or convective heating must be augmented by conversion of mechanical energy through shear, which is also needed to homogenize the product so that it expands uniformly. Specific energies for direct expansion of cereals are as high as 0.08 kWhr/kg, with high shear provided by high screw speeds and/or shallow screws.

Two types of extruder are suitable for direct expansion: the adiabatic extruder and the high-shear cooking extruder in single- or twin-screw configuration.

ADIABATIC EXTRUDERS

Originally developed for snack processing, adiabatic extrusion is a simple short-residence-time high-shear process that generates all of the heat for cooking and expansion by conversion of mechanical energy and normally uses no cooling for process control. Due to its simplicity, it is limited to a narrow range of formulas and moistures, but it can be used for some cereal products.

The short, shallow-flight screw rotates in a grooved barrel at several hundred rpm to cook and expand the product in only a few seconds. With this short time available, not all food reactions can be completed— formulations must be kept simple so that the process may be optimized for gelatinization of the basic feed grain. Furthermore, the extremely high heat flows limit the formula to materials that do not scorch easily.

The physical nature of the feed is also limited to free-flowing granular materials that enter the shallow screw easily. This limits formulation, including moisture, which should not exceed about 16%. Indeed, moisture is the primary control parameter in adiabatic extrusion, other than variable speed drive. Water may be added to the feed in two stages— in the feed hopper just before extrusion (where quick response for process control is practical) and by previous continuous or batch mixing. The latter provides enough time for the feed material to soak up the water and equilibrate, since mixing is commonly followed by a "tempering" step of up to 2 hr. This is important for product uniformity in the very short adiabatic extruder residence time. Moisture contributes to cooking and process control in a complex fashion, discussed in the next section.

Despite its limitations, the adiabatic extruder can provide the best process under some circumstances; it is simple, does not require much support equipment, and is small and inexpensive.

HIGH-SHEAR COOKING EXTRUDERS

As described in Chapter 3, high-shear extrusion consists of a series of mixing and cooking stages, utilizing energy derived from conduction, convection, and conversion of mechanical energy, optimized for a particular product. In direct expansion, mechanical energy conversion is important to create the product structure and temperature conducive to puffing and is usually most intense in the last extruder zone (just before the die), which operates like an adiabatic extruder. Since other energy inputs are also available during the much longer residence time in a high-shear extruder (typically 30 sec), the formula is not as limited, so that many ingredients may be used, and the moisture may be adjusted over a wider range to optimize product quality while maintaining expansion.

ANCILLARY PROCESSES

Dies

Die design follows the same procedure outlined for forming extrusion except that the effects of distortion are predominant in puffing (Alvarez-Martinez et al, 1988). In addition to the distortions caused by flow stresses, the product changes greatly in shape as it expands—not only in size, but in relative proportions. Under normal conditions, it does not start to expand until it is through the die and, in the case of die-face-cut products, after it has been subdivided into individual pieces. At that point, it inflates like a balloon, with similar distortion—all surfaces tend to become convex in a general "rounding" process. The degree of distortion depends mostly on the degree of expansion, but also on the viscosity and elasticity of the product.

Upon close examination of puffed products, the balloon analogy becomes evident—the outer layers of expanded pellets consist of collapsed cells, forming a skin around the more expanded interior. This skin is important in defining the product shape and can be functional in retaining crispiness longer when cereals are eaten with milk. Internal structure is also affected by elastic properties. A network of finer cells is produced when the product is more homogeneous and elastic (Miller, 1985). The structure is also influenced by die cooling and by formulation, as discussed later.

In direct expansion, shaping is constrained by both internal structure and ballooning. For expansion to occur, a minimum die-flow cross-section thickness is required to permit several layers of cells to form and create an outer skin. The actual size depends on the product and on the cell size but is about 1 mm in most cases. The assumption of a continuous, homogenous fluid loses validity when the size of the individual constituents of the product—in this case cells—is of the same magnitude as that of the overall piece. In addition, ballooning tends to prevent the creation of concave shapes. Most products made with direct expansion are simple in shape, but they can be made in more complex forms with a degree of three-dimensional shaping from flow deformations and puffing. A problem in attempting to make very complex shapes is that the product can break more easily—a particular concern in die-face cutting, where the product is exposed to violent stresses in cutting and collecting after extrusion.

Cooling

Die cooling in direct expansion adds another dimension to the formation of expanded products. In addition to the die-flow distortion discussed in the section on forming extrusion, product temperature gradients can cause differential puffing. For example, decreasing expansion at product surfaces can create an improved skin. Overall product expansion can also be fine-tuned at the die for better process control.

Cutting

Die-face cutting is the usual method of converting a directly expanded product into individual pieces. Since the product undergoes expansion for some time after emerging from the extrusion dies, the same considerations given to die-face cutting for forming apply, and the same problems exist at the die face in both cases. In direct expansion, the problem of product stickiness is not as great, since the moisture is less, but the problem of product breakage is greater. As a result, coaxial cutters are favored for their generally lower blade velocities.

Postextrusion Forming

Although limited in application, postextrusion forming can be used with directly expanded products in some cases. Since the product remains plastic for a short period of time (10–20 sec) after expansion before it cools and sets, it can be mechanically deformed to create a different shape. This is usually done by pulling the expanded strand from the extruder into special shaping/cutting machinery with rollers or conveyors. Crimp-cutting can reproduce the pillow shape of a laminated product by flattening tubular extrusions at the cutoff points. Any product compression must be done carefully, however, to avoid collapsing too much of the cellular structure and forming undesirable hard spots in the product.

Remote cutting can also be performed after the product has become stiff so that it resists collapsing. This may be done by feeding the continuous expanded product stream into a high-speed rotary cutter similar to the parallel-axis die-face cutter. The resulting product is quite different from a die-face-cut product in that it is not rounded at the cutoff faces, and those faces have open cells instead of an outer skin. When these characteristics are desired (to show cross-section shape or interesting product layers from coextrusion, for example), remote cutting can be useful. This method is generally limited to presweetened cereals.

INTERACTION OF INGREDIENTS WITH EXTRUSION PROCESSING

In addition to the natural grains used for breakfast cereal manufacturing, many other ingredients are added to improve flavor, texture, and appearance and to influence the manufacturing process. "Functional" ingredients are those that affect extrusion cooking, forming, and expansion. These may include ingredients that are not specifically added for effects on the process but that have that unavoidable consequence.

Functional Ingredients

These generally fall into four classes—sugars, lipids, salts, and inert particulates—all of which affect the physical and chemical properties of cereal products in extrusion. In high-moisture systems, sugars tend to increase viscosity at low concentrations and to decrease it at higher concentrations by inhibiting gelatinization (Bean and Osman, 1959). The overall effect on viscosity can vary with different sugars, but they all tend to increase the time required for complete gelatinization of the product. In extrusion of cereals, sugars usually tend to reduce apparent viscosity, indicated by a reduction in extruder drive-power consumption, and to demand higher cooking temperatures to achieve the same degree of product expansion. The higher temperature is needed to compensate for the retardation of gelatinization and to increase the water activity for expansion—sugars tend to tie up water needed for puffing. Higher extrusion moistures can also increase water activity at the expense of further reduction in mechanical energy conversion. Extruder screw profiles for high-sugar extrusion are selected to generate more shear and/or pressure to properly cook and expand the product.

Sugar has other effects that must be addressed. High-sugar products are stickier, requiring more careful handling, and reducing sugars can have a detrimental effect on available lysine if browning occurs (Harper, 1981). Twin-screw extrusion is particularly attractive in the latter case, since it is more controllable and reduces long residence time "tails," which can promote browning.

Salts have a similar retarding effect on expansion due to reduction in water activity. The effect on viscosity is in the opposite direction, however (Osman, 1975), leading to increased conversion of mechanical energy. To some degree these effects tend to cancel each other, so salt levels may generally be varied without changing extruder screw profiles as long as the drive motor has sufficient reserve power.

Lipids do not affect water activity, but they drastically reduce viscosity in the extruder and weaken gel strength (and therefore elasticity), even at levels below 2% (Harper, 1981). The overall effect on the process is similar to that of sugar, but with greater effect on product structure. Reduced elasticity causes coarser cell structure and less product rounding (Miller, 1985).

Even when no thermal or chemical interactions are possible, as in the case of inert materials, some physical effects may be observed in product quality. Inert particulates can affect structure by providing more nucleation centers for cell formation, resulting in a finer network of smaller cells.

Flavors

Volatile flavorings are particularly difficult to incorporate into directly expanded products. Although the natural flavors developed in cereal cooking are quite stable in puffing, flavor additives can be efficiently removed by involuntary steam distillation at the point of expansion. Not only are flavor levels reduced, but the flavor constituents are not removed uniformly, throwing the original flavoring off balance. This problem has been traditionally sidestepped by adding volatile flavors after extrusion, by a subsequent coating process such as one of those described in Chapter 7. In recent years, however, flavor companies have made progress in developing extrusion-resistant flavors. Their strategies involve flavor protection by encapsulation and special formula balancing, in addition to simply using more to begin with so that more is left after processing. The cost of some flavor loss may be offset by elimination of a separate flavoring step. Also, full dispersion of the flavor through the product may be preferred over surface coating.

Special Processes

Special processes involving more than simple extrusion are possible with direct expansion. Coextrusion has already been described in the section on forming. Two-phase extrusion is identical to coextrusion but uses a nonextruded second phase in the combining section of the die. The second phase may be a colorant to add visual contrast to

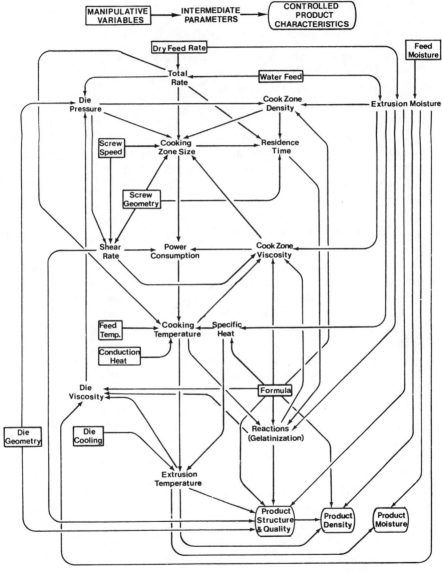

Figure 16. Interactions of process variables in high-shear cooking extrusion. Arrows point toward variables dependent on the variables from which the arrows start.

portions of the extruded piece, or a flavored material for special impact on taste. Filled extrusion is also similar to coextrusion, but it uses a liquid filling material injected into the center of a tubular extrudate. In these special processes, the injected second phase can also act as a coolant, affecting product expansion.

EXTRUSION PROCESS DYNAMICS AND CONTROL

In the foregoing discussion as well as in Chapter 3, we have discussed many formula and process variables having an effect on extruder behavior on an individual basis. We have also discussed control of the simple adiabatic and forming extruders in terms of their reactions to a limited number of parameters. The high-shear cooking extruder, especially when used for direct expansion, has more demands on its capabilities and therefore requires more freedom of control, involving a larger number of variables, which interact in a complex way.

Figure 16 is an overview of these interactions. The diagram indicates by connecting arrows which factors affect each variable. It has been simplified by elimination of convective heating (precooking and steam injection), which would affect feed moisture and temperature and overall heat input. It also departs from reality to a degree by using averages for variables that change within the process. "Cook Zone Viscosity," for example, refers to an average apparent viscosity or resistance to screw rotation in the extruder. For rigorous analysis, this should be evaluated incrementally along the screw. The average, however, is a useful concept that aids visualization of the process and is effective in determining overall process behavior and control schemes. At any rate, incremental analysis is not possible in production machines and is not a perfected method even in highly instrumented research extruders.

The diagram has been created for the most complex case but may be applied equally to simpler cases by eliminating the interactions that do not apply in a particular situation. For instance, when a variable is held constant, it does not affect process dynamics.

Upon examination of Figure 16, we find three product variables: *structure and quality* (a general, user-defined term that may be subdivided into particular qualities such as flavor and cell size), *moisture*, and *bulk density* (degree of puffing or expansion). These are indirectly controlled by 10 manipulative or independent control variables via a network of intermediate parameters.

In the product/process development exercise, all 10 of the independent variables must be considered in determining the optimum process. For control of an existing process, however, many of the independent variables are actually fixed, leaving the remainder for on-line adjustment in production. Mathematically, it is necessary that the number of control variables be equal to the number of controlled product variables. If this criterion is met, all of the product variables can be simultaneously fixed at their desired points by a unique combination of control values—assuming that the process is operating within a controllable range as set up in its development. Additional control variables may be used to broaden process conditions beyond this unique

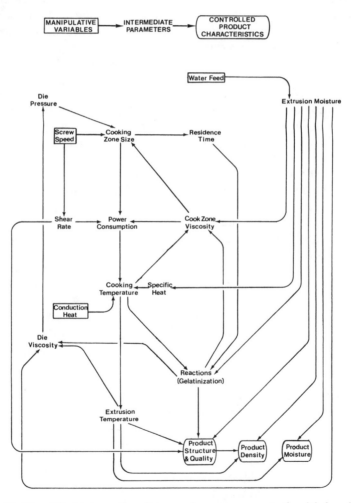

Figure 17. Simplified interaction diagram for process control, with less important and normally nonvarying parameters eliminated.

operating point. This can be useful for addressing other issues such as energy cost minimization.

The normally fixed variables include die geometry, screw geometry (except in special cases such as the APV Baker "barrel valve"), feed moisture and temperature (except where a precooker is used), and formula (except in special cases where a functional ingredient, such as a lipid, is added at a variable rate specifically for process control). Remaining for process control are dry feed rate, water feed rate, screw speed, conduction heat (or barrel temperature) and, in cases where it is used, die cooling. Feed rate is usually adjusted to satisfy another demand—production rate. Disregarding die cooling as a special case, three variables remain to control product characteristics. These are *water feed rate, barrel temperature,* and *screw speed*—the normal situation in cereal processing.

By eliminating the influences of all other variables plus that of total rate, which does not change significantly in a controlled system with constant dry feed rate, the process diagram may be simplified somewhat, as shown in Figure 17. Other simplifications, eliminating interactions of secondary importance in normal cereal extrusion, have also been made. These include the effects of die pressure on shear rate and of shear rate on cook zone viscosity, and variations in product density. With these eliminations, some variables are functions of only one other variable, so that they change together, and may be considered one. These include: cooking and extrusion temperatures; die viscosity and pressure; water feed and extrusion moisture; cook zone size and residence time; and screw speed and shear rate.

Before looking into the effects of each control variable, we should make note of the usual methods of control employed in plant processes. Normally, product characteristics are not continuously monitored as input to process controllers (although continuous methods for such characteristics as moisture and color are now available). Instead, intermediate variables, such as die pressure, power consumption, and product temperature at one or more locations, are selected for process tracking. They are intermediate because they are neither directly manipulated nor final objects of control. However, they are easy to measure continuously and, if enough of them are in the loop to match the controlled variables, they can be used to maintain constant product characteristics when controlled by the manipulative variables.

By tracing the effects of each control variable through the network of interactions, we find that the results can be contradictory, depending on which path dominates in any particular case. This is illustrated in Figures 18-20, in which the simplified effects are shown along with the direction of change, indicated by a plus sign for an increase or a minus sign for a decrease in response to an increase in the control variable. Normal interactions are indicated by solid lines, and the usually dominant effect is circled, with conflicting effects in parentheses. Effects found in undergelatinized systems are shown with dashed lines. The net results of the interactions are summarized in Table 1.

Effects of Changes in Barrel Temperature (Conductive Heating)

Barrel temperature is the most straightforward of the control variables in that it directly affects extrusion temperature, with strong influence on the amount of water vaporized as the product exits the extruder. Thus, increasing barrel temperature decreases product

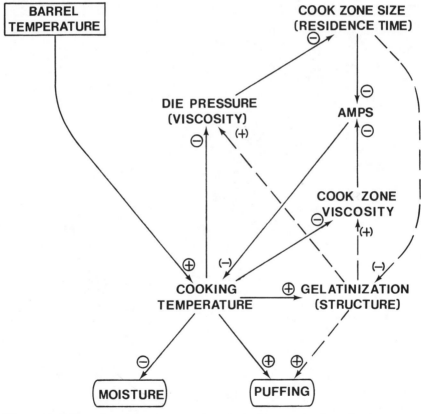

Figure 18. Effects (+ = increase, − = decrease) of increasing barrel temperature on process variables. Normal relationships shown with solid arrows. Usual dominant effect is circled; others are in parentheses. Effects of undergelatinized system shown with dashed arrows.

TABLE 1
High-Shear Cooking Extrusion: Summary of Process Interactions

Dependent Variable	Barrel Temperature	Screw Speed		Feed Water	
		Normal	Some-times	Normal	Some-times
Process temperature	+	−	+	−	+
Die pressure	−[a]	+[a]	−[a]	−	+/−
Motor power, amp	−[a]	−	+	−	+
Puffing	+	−	+	+/−	+
Product moisture	−	+	−	+	+/−

[a]In undergelatinized systems, these responses may reverse to the opposite direction.

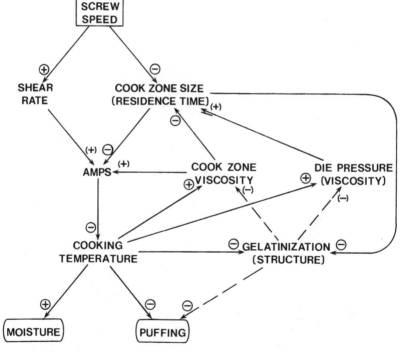

Figure 19. Effects (+ = increase, − = decrease) of increasing screw speed on process variables. Normal relationships shown with solid arrows. Usual dominant effect is circled; others are in parentheses. Effects of undergelatinized system shown with dashed arrows.

moisture and increases puffing (Figure 18). This effect is mitigated by a normal reduction in die pressure/viscosity that creates a feedback loop via cook zone size and power consumption, thereby reducing the other energy input, conversion of mechanical energy. In under-gelatinized systems, however, viscosity can increase with higher barrel temperature as gelatinization is improved, increasing the motor load and magnifying the effect of barrel temperature, with an opposite effect—or increase—in operating pressure.

Effects of Screw Speed Changes

The dominant response to increasing screw speed is a reduction in cook zone size as less extruder length is needed to pump the product, with a resulting decrease in power input and temperature. This leads to less puffing and higher product moisture. The decrease in cook zone size is magnified by the normally higher cook viscosity found at lower temperatures, through a feedback loop in Figure 19, but it is mitigated by another feedback loop created by the higher die pressure/viscosity.

In cases where the increased shear at higher screw speed is more important than the change in cook zone size, power consumption will increase, with a result opposite to the above—higher temperature with more puffing and lower product moisture at a reduced die pressure. These effects may be magnified or diminished by changes in viscosity, depending on whether gelatinization or temperature is more important. In undergelatinized systems, the effect on die pressure can reverse, as viscosity increases with better cooking.

Effects of Water Feed (Extrusion Moisture)

As in most food processes, moisture is perhaps the most important factor in extrusion. This is of particular interest for the many roles water plays: as a diluent or lubricant reducing viscosity; as a heat sink restricting temperature variations; as a reactant, particularly in gelatinization, where it tends to increase viscosity; and as a direct influence on puffing, product moisture, and structure through its effects on elasticity and plasticity. The net results on process behavior and product characteristics are complex, as might be expected.

As water feed or process moisture is increased, product moisture normally increases as well. The effect on puffing, however, depends on which dominates—extrusion moisture or temperature. Puffing increases with moisture but decreases at the lower temperatures normally found with higher moisture. This is due to the combined effects of higher specific heat and reduced viscosity in both the die and cook zones, with a resulting decrease in power consumption. As shown in Figure 20, the viscosity reduction is mitigated to some extent by the temperature reduction, which usually has the opposite effect. At higher water feed rates, we usually expect to see a reduction in power consumption, die pressure, and temperature with increased product moisture, and an indeterminate effect on puffing.

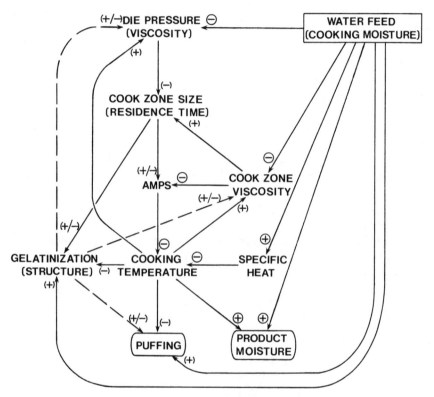

Figure 20. Effects (+ = increase, − = decrease, +/− = indeterminate) of increasing water feed rate on process variables. Normal relationships shown with solid arrows. Usual dominant effect is circled; others are in parentheses. Effects of undergelatinized system shown with dashed arrows. (Note that this variable is different from the other two because water reacts with more of the process variables, leading to contrary behavior and indeterminate results. For example, viscosity in the cook zone and the die may both be expected to drop at higher moisture. The net effect on residence time is not predictable, however, since these viscosities have opposite influences. No dominant effect can be determined, so both of these effects are indicated in parentheses, and the effect of residence time on amps is shown as indeterminate. Puffing is similarly indeterminate, but product moisture is expected to increase through three different dominant paths, including one via amps.)

In undergelatinized systems, the viscosity can actually rise with increased water. When operating in this region, power consumption, cooking temperature, and sometimes die pressure rise with moisture, resulting in increased puffing. The effect on product moisture then becomes indeterminate, depending on which is dominant—temperature or extrusion moisture.

Process Control

To illustrate how process interactions are used for actual process control, we can look at a simple but common case—control of product puffing and moisture by manipulation of feed water and barrel heating, with all other variables either fixed or allowed to float in response to changes in the control variables. We also assume the following common interactions: extrusion temperature and puffing increase with barrel temperature, while die pressure, drive motor amperage, and product moisture decrease (at constant water flow), and puffing and product moisture both increase with water flow, while extrusion temperature, pressure, and amperage decrease (at constant barrel temperature). Extrusion temperature in this context means temperature of the product within the extruder. It is important to note that both moisture and puffing are functions of both water and barrel heat,

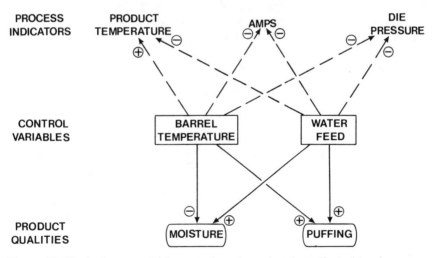

Figure 21. Typical two-variable control system, showing effects (+ = increase, − = decrease) of increase in the variables. Note that this is an example only and may not apply to all high-shear extrusion processes. Effects on process *monitors,* as shown by broken lines, may or may not be reflected in changes in the extruded product.

so they must be adjusted in tandem to control the product characteristics, as indicated in the control diagram, Figure 21.

With this simple system, manual control is feasible. To adjust for excess product moisture, the barrel heat is increased, while the water is decreased to prevent excess puffing. Excess puffing, on the other hand, may be corrected by decreasing the water while decreasing barrel temperature to maintain constant product moisture. The entire range of corrective actions may be presented in a two dimensional chart, as shown in Table 2. The amount of change needed to correct for any particular deviation from specifications must be determined experimentally. Actual control is normally accomplished by making incremental changes.

Two-variable systems are easy to visualize and lend themselves to manual control as described. As more variables are added to the control system, however, the situation becomes more complex. Three-variable control, for example, requires a three dimensional chart, and four variables exceed human abilities for visualization. In these cases, computer control, utilizing an experimentally determined mathematical model, is useful. The computer can "visualize" any number of dimensions.

Even in our simple case, automatic control can be applied. This can be accomplished by simple controllers with some human intervention or with a more sophisticated system. For simple control (not possible in all cases), we may use the easy-to-measure intermediate variables of die temperature and pressure, which vary predictably with other changes. Barrel temperature or heat flow can be controlled by a simple proportional controller from a product temperature signal; heat flow is increased if product temperature drops below a set point. Water

TABLE 2
Typical Two-Variable Control Strategy for High-Shear Extrusion Processes[a]

	Low Product Moisture	Moisture OK	High Product Moisture
Insufficient puffing	Increase water	Increase barrel temperature and increase water	Increase barrel temperature
Puffing OK	Decrease barrel temperature and increase water	No change	Increase barrel temperature and decrease water
Excess puffing	Decrease barrel temperature	Decrease barrel temperature and decrease water	Decrease water

[a]This chart assumes the responses shown in Figure 20.

flow can be similarly controlled by the pressure signal; water flow is increased if pressure rises. The extruder may be expected to operate consistently and produce a uniform product if these are kept constant.

Since we are not directly controlling product characteristics by this method, however, periodic product measurements must be made to calibrate the process by changing controller set points to correct for product variations. In our example, high product moisture is corrected by increasing the temperature set point. This action will also cause a decrease in die pressure, which will automatically be corrected for by a decrease in water flow. The net effect on puffing will be slight, since greater temperature and less water flow tend to cancel each other. Excess puffing can be corrected by increasing the pressure set point, which decreases water flow and also causes the temperature to rise, with a resulting automatic decrease in heat input. The effect on product moisture should not be great. After manual adjustment of the controllers is made, the process can usually run by itself for a period of time until the next product check is made.

Although the example given above is typical, it is not universal. For any actual process, the interactions must be determined experimentally and then applied to a control system. It is most important to determine which product characteristics are essential, in any case, and then to determine the critical control variables to which they respond most significantly.

Scaleup

We have already discussed extruder scaleup for cases where heat transfer by conduction is a significant energy input, as in forming or low-shear cooking extrusion. The same principles apply in the high-shear cooking extruder, but with less emphasis on heat transfer, which is usually a minor energy source—and one that is not used at all in adiabatic extrusion.

With heat transfer eliminated, the scaleup becomes primarily dependent on volume—production rate increases with the cube of the screw diameter for geometrically similar machines. The previous arguments were based on the assumption of ideal fluid flow, however, which is an acceptable approximation in high-moisture systems dominated by heat transfer. In high-shear extrusion, on the other hand, the typical non-Newtonian behavior can be significant. Cereal products are shear-thinning, or pseudoplastic. In practical terms, this means that the product does not shear uniformly, but more at the machine surfaces and less in the middle of the screw channel. Shear, responsible for heat generation, then becomes to some degree a surface phenomenon like heat transfer.

For single-screw machines, the situation is often addressed by varying machine proportions as overall size is increased. This leads to relatively shallower screws operating at reduced rpm in larger machines to duplicate as closely as possible the environment of shear or the product cooking conditions of the smaller extruder. The net effect of these changes is that the scaleup factor is somewhat less than the cube of the diameter that one expects with ideal fluids.

Twin-screw extrusion is less sensitive to nonideal fluid behavior because its shear rate is more uniform throughout the screw channel and the product does not stagnate. Accordingly, these machines are usually scaled up with geometric similarity and operate at the same rpm over a wide range of sizes, closely following the cube rule for production rate.

References

Alvarez-Martinez, L., Kondury, K. P., and Harper, J. M. 1988. A general model for expansion of extruded products. J. Food Sci. 53:609-615.

Bean, F. J., and Osman, E. M. 1959. Behavior of starch during food preparation. II. Effects of different sugars on the viscosity and gel strength of starch pastes. Food Res. 24:665-671.

Bernhardt, E. C. 1967. Processing of Thermoplastic Materials. Reinhold Publ. Corp., New York.

Donnelly, B. J. 1982. Teflon and non-Teflon lined dies: Effect on spaghetti quality. J. Food Sci. 47:1055-1058, 1069.

Harper, J. M. 1981. Extrusion of Foods. CRC Press, Inc., Boca Raton, FL.

Harper, J. M. 1986. Extrusion texturization of foods. Food Technol. 40(3):70-76.

Hawkins, M. D., and Morgan, R. G. 1986. Similitude approach to food extrusion die flow problems. ASAE Paper No. 86-6001. Am. Soc. Agric. Eng., St. Joseph, MI.

Janssen, L. P. B. M. 1978. Twin Screw Extrusion. Elsevier Scientific Publ. Corp., New York.

McKown, W. L., Zietlow, P. K., and Ball, M. K. 1969. Process for making breakfast cereal. U.S. patent 3,464,826.

Miller, R. C. 1985. Low moisture extrusion: Effects of cooking moisture on product characteristics. J. Food Sci. 50:249-253.

Osman, E. M. 1975. Interaction of starch with other components of food systems. Food Technol. 29(4):30-35, 44.

Padmanabhan, M., and Bhattacharya, M. 1989. Analysis of pressure drop in extruder dies. J. Food Sci. 54:709-713.

Rossen, J. L., and Miller, R. C. 1973. Food extrusion. Food Technol. 27(8):46.

Rossen, J. L., and Miller, R. C. 1974. Method of producing laminated comestible products. U.S. patent 3,851,084.

Tadmor, Z., and Klein, I. 1970. Engineering Principles of Plasticating Extrusion. Van Nostrand Reinhold Co., New York.

Tayeb, J., Vergnes, B., and Della Valle, G. 1988. Theoretical computation of the isothermal flow through the reverse screw element of a twin screw extrusion cooker. J. Food Sci. 53:616-625.

Wilkinson, W. L. 1960. Non-Newtonian Fluids. Pergamon Press, New York.

Chapter 7

Application of Nutritional and Flavoring/Sweetening Coatings

ROBERT E. BURNS
ROBERT B. FAST

In the early days of the breakfast cereal industry, pioneer manufacturers depended on consumers to liven up the flavor of cereal in their bowls by topping off their flakes, granules, shreds, or oatmeal with sugar or honey and more sophisticated flavor and texture additives, such as blueberries, nuts, or maple syrup. The list was limited only by the consumer's imagination. However, it was not long before these variations gave birth to prepackaged product extensions, starting with presweetened flakes or puffs and continuing with more complicated products, such as fruit-filled shredded grain biscuits. Also, the need to nutritionally fortify cereals soon arose, as consumers began to view cereals as a complete meal rather than part of a well-rounded breakfast including eggs, meat, and juice. Further complicating the process in some cases is the need to add water to flake products or remove it from them to maintain a proper moisture level and thus ensure good texture and freshness over an extended period of time.

As new-generation cereal products were developed and the demand for them increased, new production techniques evolved. In many cases these new-generation cereals were and are made using an existing product as the base (flakes, puffed grains, or shredded biscuits). As coating systems were added to existing production lines to produce consistently and uniformly coated cereals, application techniques that worked effectively at flow rates of 10–20 lb/min were found to be inefficient at flow rates of 80–100 lb/min.

Flavors and additives such as vitamin powders incorporated prior to processing are subject to deterioration due to the high pressures and temperatures in processing the cereal itself. If they are to be incorporated before processing, it is therefore necessary to oversupply flavor and nutritional additives or furnish additives capable of withstanding these processing conditions. In either case the net effects are higher production costs and inconsistent quality. If the process permits, flavors and additives may be applied after extrusion, flaking, or baking to bypass these intense conditions and thus avoid losing their initial flavor and effectiveness. Additionally, processing and ingredient costs can be more strictly controlled by eliminating both the overapplication of additives and the varying rate of loss during processing. Flavors and sugars applied topically also dissolve and disperse more quickly and thoroughly in the milk than additives mixed within the cereal piece, thereby presenting a more intense flavor profile.

In this chapter, we describe and explain a number of techniques and details of coating processes for applying sweeteners and other flavors and nutritional fortification to breakfast cereals.

The Process

Coating system, flavor system, wetting reel, and *enrober* are a few examples of the nomenclature used to describe the portion of the production line dedicated to the application of liquid or dry additives to a base product. To better isolate what occurs at this stage of production, we refer to it as the *coating process.*

There are at least two separate and distinct phases of almost every coating process, as shown diagrammatically in Figure 1. The first of these (*Phase I*) may control the nutritional value, texture, and shelf life of the product, but the second (*Phase II*) involves coatings of sugar or predominant flavors and is more obvious to the consumer. Additionally, flavor bits, such as nuts, coconut, or dried fruit, can be added independently of the liquids to produce a composite or tack-on product. Cereals such as cornflakes, puffed grains, and other nonpresweetened products are subject only to Phase I, in contrast to presweetened cereals, which almost always involve both phases.

Regardless of the method of application or type of coating being applied, it is critical that the feed rate of uncoated cereal passing into the coating zone be precisely controlled. In some cases, the consistent output of an extruder may provide the necessary control, but even then a surge feeder is desirable. Ideally, inferior cereal pieces and fines or dust should be removed by screening before the cereal is fed into the surge hopper or reaches the coating zone.

The actual feed rate may be controlled either volumetrically or gravimetrically. With volumetric control (Figure 2), the product bed depth on a vibratory tray or conveyor belt and the speed of the tray or belt are the variables to be adjusted, but variations in the feed rate can occur as a result of changes in the cereal bulk density as well as operator error. A gravimetric feeder eliminates the first of these sources of variation in the feed rate.

Phase I

The additives applied in Phase I vary with the requirements of the process as well as of the finished product. For example, in the past, vitamin fortification ingredients were added prior to extrusion, milling, or flaking, but the waste described previously indicates the need for more efficient methods of applying these additives to the process stream. It is critical that additives, like any component of a product, be applied consistently and uniformly. Operationally, one can dump all the ingredients into a single batch and depend on a blender to mix them together over an extended period of time prior to milling or extrusion. This procedure is easier than the exacting task of distributing a few ounces of liquid over 100 lb or more of cereal in only a few seconds.

The following discussion details the various options for vitamin fortification during Phase I. Apart from the actual preparation and storage of the vitamin suspension as detailed in Chapter 10, the same

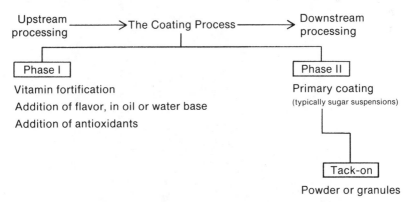

Figure 1. The coating process. Upstream processing refers to puffing, shredding, flaking, toasting, and other forming processes; downstream processing refers to drying, cooling, blending, and packaging. Flavoring and antioxidants may be applied before or following the Phase II coating instead of in Phase I. The tack-on coating with powder or granules may include powdered sugar or honey, or flavor bits such as nuts or dried fruits.

principles apply to virtually all Phase I additives. Vitamin fortification is available as a preblended ready-to-use suspension or as a liquid or powder concentrate to be mixed by the processor. Depending on the concentration, the typical target application rate can vary from 0.15 to 2.0%, the target rate being the amount specified to be applied per pound of the base product. The effective application rate is subject to variations as the mass or volume of the cereal fluctuates at the coating zone. It can be varied by either a manual adjustment by the operator or, in cases where weigh belts and microprocessor controls are utilized, an automatic adjustment.

The point of application, or coating zone, is the area where the liquid or dry ingredients are added to the base product. Regardless of the total length of the process conveyor or drum, the coating zone is the specific area of application. Any area beyond this point is for the purpose of blending in the additives applied or transferring the product to the next processing stage.

PHASE I CONVEYOR-BELT APPLICATION

For a number of reasons it is sometimes necessary to apply vitamin

Figure 2. Typical low-level volumetric vibratory feeder below surge hopper. (Courtesy Spray Dynamics, Costa Mesa, CA)

fortification to a product as it travels on a conveyor, as opposed to the more efficient coating drum method. Whether because of the cost of adding a drum to the process, space restrictions, or the fragility of a product incapable of accepting additional handling, a conveyor is sometimes the only practical option. The following guidelines for conveyor-belt processes can greatly improve the consistency and uniformity of the finished product.

Spray System

The spray system should be designed to apply the vitamin solution over the entire width of the processing conveyor, regardless of variations in the application rate. Systems for this purpose are described later. Continuous recirculation of the vitamins is required, to prevent the suspension from stratifying in either the supply tank or the process piping. Other additives may require control of the fluid temperature as well as the addition of a mixer to the supply tank to achieve the necessary fluid condition.

Product Bed

The product bed must be as thin as practically possible. The nature of the process requires that a sufficient amount of vitamins be applied to coat not only the pieces on top of the product bed but also those beneath. If not enough is applied, the top layer becomes saturated with vitamins, and the pieces beneath receive little, if any (Figure 3). At a low rate of application, the vitamin solution is easily absorbed by

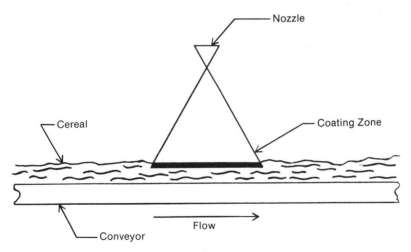

Figure 3. The coating zone in belt application.

the top layer, and the transfer of excess coating as cereal pieces rub against each other downline is minimized. In the finished product, consistency is usually a result of the intermixing of the saturated pieces with the uncoated pieces as much as it is a result of a truly uniform application. Intermixing may provide adequate consistency in large samples, but it results in wide variations in small samples and even greater variations from piece to piece. Clearly this is the primary restriction of this process.

The product bed depth can be decreased by two methods, used in combination or independently. First, it can be reduced by widening the product stream at the point of application, by means of an oscillating rake or a vibratory transfer conveyor feeding onto the processing belt (Figure 4). Second, it can be further reduced by increasing the speed of the processing conveyor relative to the infeed and take-away conveyors (Figure 5). As the product transfers onto the processing conveyor, it thins out, and as it transfers onto the take-away conveyor, it restacks. The ultimate (though impractical) objective would be a monolayer of cereal pieces. Anything approaching this contributes to the uniformity of the application.

PHASE I LIQUID SPRAY SYSTEMS

Many spray system configurations are in use around the world. Despite this variety, there are two basic principles that allow the successful distribution of a small volume of liquid over a product bed area—pressure and interference or force.

Pressure

A spray system with restrictive-orifice nozzles, used in conjunction with a positive-displacement pump, illustrated in Figure 6, is a common method of application. A single pump controls the application rate, and the size and shape of the opening in the nozzle dictate the size and shape of the spray pattern. The orifice restricts the flow of liquid, creating back pressures of 60–120 psi. As the liquid is forced through the nozzle at this high pressure, it expands into a spray. A slit opening creates a fan-shaped spray; a circular opening creates a conical pattern.

Problems arise when the application rate decreases. This change in the application rate causes the pressure at the orifice to drop, and consequently the size of the spray pattern is reduced, potentially to a stream and not a spray at all. Conversely, as the application rate increases, overpressurization creates an airborne mist, or overspray,

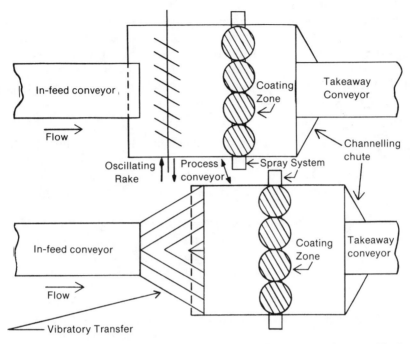

Figure 4. Decreasing the product bed depth by widening it with an oscillating rake (top) or a vibratory transfer conveyor (bottom). The latter may be grooved to assist in spreading the product across the process conveyor.

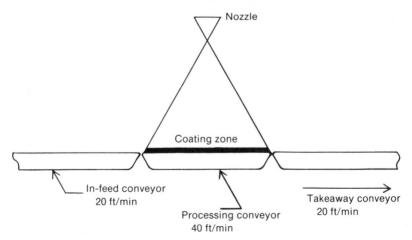

Figure 5. Decreasing the product bed depth by increasing the processing conveyor speed relative to the speeds of the infeed and take-away conveyors.

which contaminates the surrounding area and causes the product to be under-coated.

With all nozzles being supplied by a single pump, any pressure drop from nozzle to nozzle adversely affects the system's ability to apply an even layer of liquid to the product bed. The dead-end nature of this type of system can create a settling-out of powdered additives, which can potentially clog relatively small nozzle openings, compounding the uniformity problem.

Interference or Force

Compressed air (15–45 psi) can be combined with the liquid before or immediately after the liquid is discharged from the nozzle. Altering the pressure of the air affects the droplet size of the liquid; an increase in pressure disperses it over a larger area, creating finer droplets and potentially a mist.

In addition to the problems associated with single-pump dead-end systems, there are other disadvantages of this approach. Light liquids, such as oil or water, tend to mist easily, and even when the air is well controlled, the system tends to create mist and costly overspray. Typically this is restricted by using only a minimal amount of air, but then the system fails to achieve uniform coverage on the product bed. Moreover, changes in the application rate require manual adjustments in the airflow to maintain consistency; failure to make these adjustments contributes to uneven coverage. The use of plant air can also pose potential contamination problems, since it normally contains entrapped compressed oil and water droplets as well as other contaminants, which can be difficult and expensive to filter out.

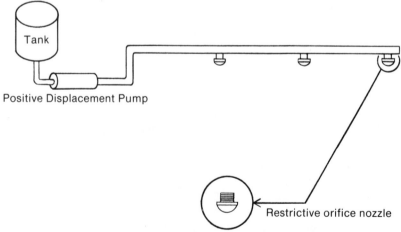

Figure 6. Restrictive-orifice nozzle spray system.

Combination System

Consistency and uniformity are ensured by combining the most effective features of a positive-displacement pump with those of a self-adjusting spray system (Figure 7), thereby minimizing the problems inherent in conveyor-belt application and low-volume spray systems in general. The combination system consists of a series of adjustable-volume positive-displacement metering pumps, each mounted on a pump-supply manifold (Figure 8). The pumps draw from a manifold containing a recirculating flow of the micronutrients to be added, which ensures that the fluid remains consistent. The number of pumps required depends on the process parameters, such as the width of the conveyor and the rate of application.

Each pump has a corresponding poppet nozzle either mounted directly on the pump or remote-mounted in the case of drum application. The nozzle is equipped with a spring-loaded poppet head (Figure 9), which regulates the pressure of the liquid at the point of discharge, ensuring that the liquid is distributed over a surface of approximately 75–100 in.^{2}, regardless of changes in the application rate. As discussed earlier, it is the pressure of the liquid along with the configuration of the orifice that creates the spray action.

The pressure of 95 psi required to expand the poppet creating the orifice is formed as the piston in the pump travels forward, displacing a precise volume of liquid. This liquid is channelled through the pump body toward the poppet, which seals the orifice.

The pressure of the liquid compresses the spring in the poppet, causing it to extend and spray the liquid out in a 360° pattern. Each actuation or cycle of the pump displaces between 0.0001 and 0.10 oz of liquid. As the volume of liquid displaced per actuation increases, the poppet is correspondingly forced further open, enabling the pressure to remain constant and in effect acting like a self-adjusting orifice. The nozzle actuates 70–150 times per minute, with the frequency of actuation depending on the application rate. Any variation in the cereal production rate can be quickly accommodated by incrementally increasing or decreasing the actuation frequency. The air used to drive the piston never comes in contact with the liquid or the cereal and is controlled by a self-contained lubricating and regulating pneumatic control console.

With poppet nozzles, unlike restrictive-orifice nozzles, no time is required to build up pressure, so that every pulsing spray is full and complete. Just as important, the spring-loaded poppet provides a positive shutoff, eliminating the drips and runs common with the other systems.

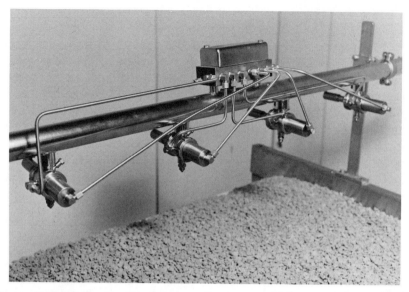

Figure 7. Self-adjusting liquid spray system (as detailed in Figures 8 and 9) in actual operation over a belt conveyor. (Photo of Meter Master system courtesy of Spray Dynamics, Costa Mesa, CA)

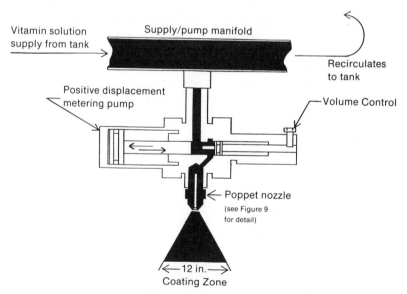

Figure 8. Sectional view of positive-displacement/metering spray station with self-adjusting nozzle.

PHASE I COATING-BLENDING DRUM APPLICATION

If the product can be tumbled, a coating drum, such as the one shown in Figure 10, is the most efficient way to apply both Phase I and Phase II coatings. A coating drum may act both as a blender and as a method of exposing the cereal to the spray, thereby producing an immediate and complete coating.

The coating of vitamins in a Phase I system can be added at either point 1 or point 2 in Figure 11. In both cases the same problems that occur in the conveyor process also occur in the coating drum. The coating zone is small relative to the volume of cereal going into the drum, which means that sufficient solution must be applied at either point to fortify all the cereal below and on either side of the coating zone. Once again, the cereal directly in the coating zone becomes supersaturated with vitamins, whereas the rest receives little or no coating. Over 80% of the coating can end up on less than 20% of the cereal.

The same poppet or restrictive-orifice spray system or something as simple as an open-ended pipe with no nozzle can be used to apply the vitamins at either point. The area of coverage or application is

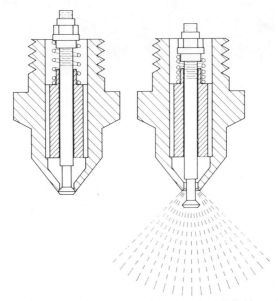

Figure 9. Enlarged section of poppet nozzle tip, closed (left) and open (right); rate of actuations, 70–150/min.

less critical than the drum's ability to blend the saturated cereal pieces throughout the product stream. In this case the drum is used strictly as an in-line blender. Its immediate turbulent blending action allows for a marginal degree of rub-off as the over-coated pieces come in contact with the uncoated pieces, transferring at least some of the excess coating. However, in any process where the application rate is minute or the cereal is absorbent, the likelihood of transferring much coating in this manner is minimal.

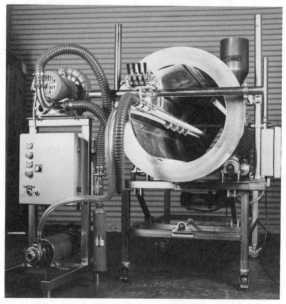

Figure 10. Coating drum with sugar spray system and powder-granule applicatior mounted. (Courtesy Spray Dynamics, Costa Mesa, CA)

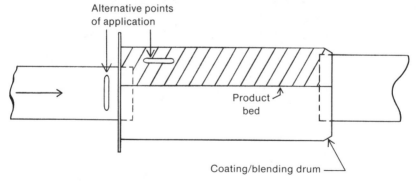

Figure 11. Alternative points of application of Phase I (vitamin) coating by the drum method.

The greatest obstacle is having insufficient dwell time within the drum to achieve the proper degree of blending. At flow rates of 80 lb/min, even 30 sec of blending requires a drum over 10 ft in length. Even so, all that is achieved is a consistency based on the uniform distribution of a few saturated pieces in the entire product stream.

Space restrictions often result in the use of a drum that is incapable of providing the proper blending action. In this case, as with the conveyor, the downstream movement of the product is counted on to provide additional blending. In some cases an oversupply of flavors and additives is used in the hope of compensating for poor blending, but this results in increased production costs and even wider variations in product consistency.

The degree and duration of blending can be greatly reduced if the vitamin suspension is applied over a larger portion of the product bed. With a liquid application system similar to that already described, the same volume of liquid dispensed over 4 in.^{2} can be applied over nearly 200 in.^{2}, coating the product almost instantaneously within the first 3 ft of the drum, assuming a production rate of 80–100 lb/min (Figure 12). The system is basically the same as the conveyor process, except that the poppet nozzles are remotely mounted on a spray arm within the drum. The spray arm serves as a jacket for the internal stainless steel lines that supply each of the nozzles. For solutions with a high melting point, it can be hot-water-jacketed to maintain the proper temperature up to the point at which the fluid is sprayed. Once again, as in conveyor processing systems, the number of actuations per minute is increased or decreased to adjust for fluctuations in the cereal production rate.

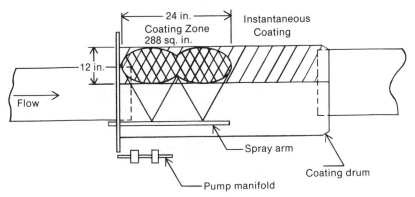

Figure 12. Phase I spray coating, showing large area of application achieved by a spray arm mounted within a coating drum.

Coating Drum

Uniformity is achieved by matching the large coating zone with a drum capable of exposing all of the flakes to the spray within the coating zone. The drum must be designed to create a folding action that carries the bottom layer of the product bed from the six o'clock to the nine o'clock position, from which it cascades down the surface of the bed. The fine spray hits the cereal pieces as each is exposed to the spray, as shown in Figure 13. Even at 10–14 rpm the cereal pieces must always fall onto the cereal bed rather than the drum surface; the cereal bed provides a cushioned landing that enables even the most fragile products to be processed.

The height and angle of the flights vary according to the bulk density of the cereal. If the lift is too great, cereal pieces can exit the coating zone without being exposed to the spray, and the force of the drop back down to the drum may break them. Conversely, flights with too

Figure 13. Folding action carries the bottom layer of the product bed from the six o'clock to the nine o'clock position and exposes each cereal piece to the spray (see Figures 10 and 12 for drum system details). (Courtesy Spray Dynamics, Costa Mesa, CA)

low an angle or height fail to create the required lift to generate an efficient folding action, and the product bed can pass through the drum with virtually no tumbling or blending. Even so, a certain degree of tumbling occurs as the top layer of cereal is coated; the coating increases the bulk density of the cereal, thus creating a blending action as the heavier, coated pieces work their way to the bottom of the product bed.

The angle of the drum must also be adjustable (3–8° from the horizontal) to maintain a product bed depth (a maximum of 6–9 in.) and width (16–24 in.) that allow the folding action to bring most of the cereal pieces to the surface of the bed within the coating zone. The retention and infeed flanges contribute to maintaining a consistent product bed throughout the length of the drum, ensuring that any applied liquid or dry ingredients hit the cereal and not the drum surface.

Changes in the parameters of the process may require adjustments in the size of the coating zone and the size of the drum. Higher production rates and corresponding decreases in the dwell time of the cereal within the drum require a larger coating zone, which can be created by adding more metering pumps and nozzles. The drum may also need to be increased in size for multistage applications. In some cases, such as the application of high levels of sugar to puffed products, the flights may be removed in the Phase II coating zone to minimize buildup in the drum, with the folding action then being generated entirely by the increased bulk density of the coated cereal pieces.

Whether the Phase I liquids are applied using a conveyor or a drum system, the increase in moisture must be within the range of acceptable total moisture for the finished product, to avoid the need for an additional redrying step. Redrying can usually be avoided by spraying the hot product as it exits the toasting or baking step; this allows the added moisture to flash off as the product cools.

Phase II

The types of coating applied in Phase II are various blends of sugar and water or sugar and fruit juice; honey; supersaturated sugar slurries; and custom coatings that also contain oils and other flavorings. Some details of formulation are discussed later in this chapter.

PHASE II CONVEYOR-BELT PROCESS

A conveyor may be the only means of moving the base product through the coating zone, because of the nature of the process and the handling characteristics of the product. In some cases a conveyor is the only means suitable for a product coated on a single side, such as frosted biscuits or granola bars. The most effective and efficient method of coating any product comprehensively is a coating drum system. But if a product is to be coated on one side only or if for any reason it cannot go through a coating drum, a conveyor-belt system may be the only feasible method, and the guidelines detailed earlier should be followed.

Because of the importance of the visibility of the coating on the finished product and the high application rates (up to 50% or more), uniform application is absolutely critical. Buildup of the coating on the conveyor as a result of overspray or gaps in the product flow are significant when 1,000 lb or more of sugar per hour is being sprayed over a belt 48 in. (122 cm) wide. The foundation of any conveyor-belt application is that a given amount of coating is applied for every square inch of the conveyor entering the coating zone, whether the product is there or not.

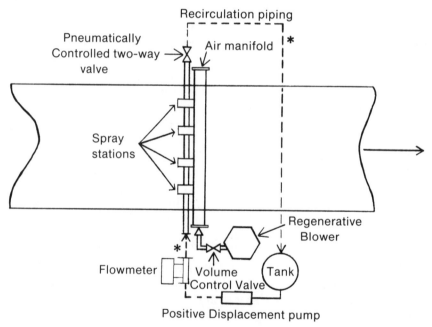

Figure 14. Phase II (sugar suspension) spray system for conveyor belt with spray stations supported by common product and air manifolds, product pump, and regenerative air blower. Note the two-way valve enabling the system to operate in a dead-end mode. The diameter of the piping depends on viscosity of the spray suspension, flow rates, and piping layout. A heat exchanger may be added at points indicated by an asterisk.

A dead-end system with restrictive-orifice spray nozzles could be utilized, but the tacky nature of sugar solutions combined with undissolved sugar particles, flavoring, and caramelized sugar causes the nozzles to plug frequently, and thus gaps in coverage are created as the cereal passes through the coating zone. Additionally, as one nozzle becomes partially obstructed or clogged, the additional back pressure in the line forces the solution through the remaining nozzles, and the uniformity of application is further affected.

The nozzle orifices can be enlarged, but this may not enable the pressure to build sufficiently to create a spray pattern capable of covering the required coating zone. As a result, designers of systems based on this approach try to strike a balance between proper orifice sizing to create the desired spray pattern and the need to frequently change nozzles as they become clogged or partially obstructed.

A sugar spray system that does not clog and has the capability of spraying all of the primary coatings discussed thus far can be designed for use over a conveyor belt. It has the additional advantage of spraying suspensions with particulates up to 3/16 in. in diameter. Such a system is illustrated in Figure 14. More than just holes in a pipe, this system consists of a series of spray stations supported by a common product-air manifold, product pump, regenerative air blower, and control console. The number of spray stations and their configuration depend on the process parameters. Because of the difficulty of spraying more viscous solutions, the nozzles may need to be spaced quite close together (see the photograph of an actual installation in Figure 15).

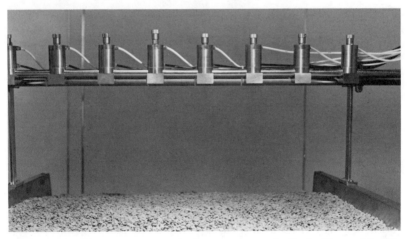

Figure 15. Actual spray system, as shown schematically in Figure 14. (Photo of Meter Master Clog-Free system courtesy of Spray Dynamics, Costa Mesa, CA)

A top- and bottom-enrobing system can be constructed by positioning manifolds above and below a conveyor belt housed within a cabinet. The greatest difficulty in this is compensating for the areas deprived of coating where the belt blocks the spray from the bottom manifold. Nozzles vary from 1/32 to 1/4 in. in diameter. The flow rate per spray station depends on both the size of the orifices and the pressure of the fluid within the product manifold. Nozzle sizing is determined by both the maximum particulate size and the application rate.

Variations in the spray rate from one side of the manifold to the other due to a decrease in pressure across the manifold are corrected by means of adjustable orifices (Figures 16 and 17). All the orifices are of equal size, but by means of an adjustable piston, a partial restriction can be created, equalizing the pressure and therefore the flow rate across the coating zone. Once set, the nozzles need to be adjusted only as the operational parameters change significantly. The

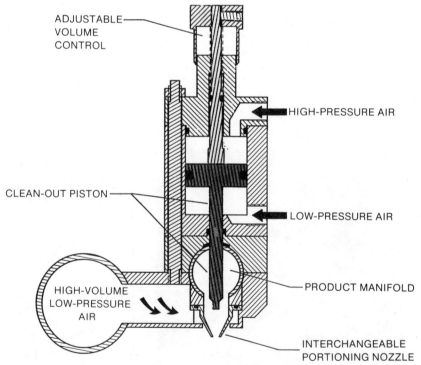

Figure 16. Section of Clog-Free spray station with adjustable orifice, as shown in operation in Figure 15. The adjustable volume control works by restricting the nozzle opening. The cleanout piston fires through the nozzle at up to 99 actuations per minute with a force of 860 psi.

same piston that restricts each orifice also serves as a shutoff, providing positive shutoff when the system is not operational and during on-line cleaning-in-place. Even more importantly, this piston provides continuous clog-free operation by firing between one and 99 times per minute. The pulse itself takes only a fraction of a second. A pressure of 800 psi at the orifice forces out any obstruction (nuts or other particulates or sugar buildup), ensuring continuous spray action. As long as the orifice and the pump rate are constant, the system sprays a consistent, uniform pattern across the belt (Hanson, 1981; Mancuso, 1983). The components of this system and its operational sequence are detailed in the following section of this chapter, on drum application of Phase II coatings.

In addition to the above-described spray nozzle application methods, spinning disks have also been used. In this method, a disk 4–6 in. in diameter is directly coupled to a high-speed motor capable of 3,600 rpm or more. The disk shape is dictated by the flow characteristics of the material to be sprayed, the desired droplet size in the spray, and the area to be covered.

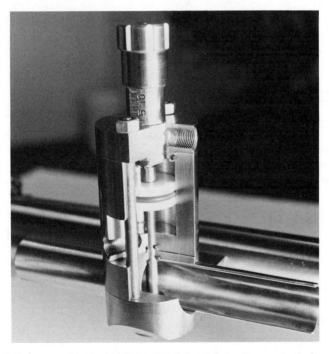

Figure 17. Spray station, as in Figure 15, cut away to show patented mechanism (Hanson, 1981; Mancuso, 1983). (Courtesy Spray Dynamics, Costa Mesa, CA)

The disk method has the advantage of operating without clogging. Since the material to be sprayed is metered to the disk from a metering pump through a pipe of relatively large diameter (e.g., 1/8–3/8 in.), spray stoppages due to clogging are unlikely to occur.

The disadvantage of the disk method is the difficulty of spraying in a confined pattern, such as downward onto a product moving along a belt. This may be controlled to a degree by the position of the feed tube in the disk, but the spray is still produced in the full 360° around the disk.

A caution to be observed in all conveyor-belt applications is the need to keep the belt clean, particularly in sugarcoating applications. As previously noted, the application of sugar is typically continuous, whether or not the product is present on the belt within the coating zone. Whether the belt is solid neoprene or wire mesh, it is best to include a belt-washing device on the return portion of the belt cycle to keep it clean.

In the case of a solid belt, the washer can be a reservoir of recirculating heated water in which a set of rotating brushes continuously cleans the belt as it passes through. A set of wipers removes the excess water as the belt exits the washer. A regenerative air blower can provide air jets or a stream of warm air to assist in drying the belting. In the case of wire mesh belting, sprays of water can be used in place of (or along with) the washing brushes.

PHASE II COATING-BLENDING DRUM APPLICATION

As previously discussed, a coating drum can be used either for blending or to expose the cereal to a large coating zone. Applications of heavy sugar solutions certainly benefit even more from the latter use.

If the sugarcoating is added in a localized area, as in Figure 18, and the drum is depended upon to distribute the coating through the remainder of the product, two things happen. First, as the product bed tumbles over, the mass of sugarcoating hits the surface of the drum, creating an accelerated buildup problem. The sugarcoating material is heavier than the cereal and therefore stays between the cereal pieces and the wall of the drum. Over time the cereal that builds up on the drum becomes supersaturated and either gets scraped off or falls into the product stream as an unwanted contaminant in the cereal carton. When dried, such sugar balls become extremely hard and may be similar in size to the cereal pieces, so that they are difficult to remove by screening. The second problem, as outlined earlier, is poor distribution of the sugarcoating throughout the product stream. Visually this variation is difficult to detect, because of the high rate of application. As the product travels through the drum, the rub-off effect coupled with the coating material on the drum itself does cause the coating to spread around, but the variation can still be significant. This variation can prevent the adherence of flavor bits or nuts added separately after the liquid coatings are added.

The nonclogging spray system described earlier can be utilized in conjunction with a coating drum (Figure 19) and practically eliminates both of these problems. With a liquid coating zone of 400 in.^{2} or more, the amount of coating that is applied is only what is needed to coat the surface layer of the product. As previously discussed, this approach eliminates the need for overapplication and excessive downstream blending. Once the cereal exits the coating zone, most of the cereal pieces have the specified level of coating.

If the suspension contains particulates, such as nuts, or a component that tends to settle out, such as cinnamon, the spray system can be operated in a recirculating mode. In this case each spray station draws off a portion of the flow going through the product manifold. The flow rate of the pump is sufficient to supply each spray station and provide enough additional volume to recirculate the remaining liquid back to the supply tank (Figure 20). This recirculating action prevents the particulates from settling out within the piping.

Figure 18. Phase II (sugar) coating applied in a localized area of the drum. (Courtesy Spray Dynamics, Costa Mesa, CA)

More exacting control can be obtained by operating the system in a dead-end mode. A pneumatically controlled two-way valve positioned after the last spray station (see Figure 14) ensures that the exact amount of sugar solution pumped by the positive-displacement pump is forced through the nozzles. A flowmeter can be positioned between the pump and the spray station manifold, allowing the system to be monitored and controlled by a central control station. In most cases the consistency of a positive-displacement pump combined with a programmable speed controller is more than sufficient to ensure precise application.

In either a recirculating or a dead-end system, the pressure of the liquid at the point of discharge is 5–15 psi, which is much lower than the 60–150 psi in a restrictive-orifice nozzle. The blower supplies high-volume, low-pressure air (30–50 cfm and 0.75–2.5 psi) that surrounds the stream of liquid, creating a simultaneous interference and

Figure 19. Nonclogging spray system applies Phase II coating over 400 in.^{2} or more of folding product bed in a coating drum. (Courtesy Spray Dynamics, Costa Mesa, CA)

accelerating effect that breaks the stream into a solid cone spray. A manual dump valve on the regenerative air blower allows the operator to control the spray pattern by regulating the volume of air. The stacking effect within the air manifold creates equal air pressure at each spray station.

Spray systems using plant air are common throughout the industry. One of the primary reasons plant air is used is its accessibility. Unfortunately, the high cost of manufacturing compressed air combined with the need to clear factory air of contaminants by an elaborate filter system makes compressed air an expensive and inconsistent component of any spray system. The low-volume, high-pressure air (2–8 cfm and up to 80 psi) also overaccelerates the fluid, creating costly overspray and mist. The low-velocity approach just described provides the greatest flexibility in adjusting the size of the spray pattern.

Dry Flavor Bit Application

Dry flavor bits such as nuts, dried fruit bits, coconut, or powdered honey can be adhered to the cereal pieces within the coating drum immediately after the Phase II (sugar) coating is applied. The coating serves as the tack to adhere the flavor bits to the cereal.

Figure 20. Twin-tank mixing and supply system with self-contained hot water temperature-maintenance system. Size and configuration of tanks are dependent on process parameters. (Courtesy Spray Dynamics, Costa Mesa, CA)

The size and relative positioning of the flavor bit coating zone within the drum is dependent on the application and overall production rate. A vibratory feeder or a powder-granule dispenser (Figure 21) can apply flavor bits uniformly over a 2 × 36 in. coating zone (Hanson, 1985).

In conveyor-belt applications, such particulate matter can be applied via the same powder-granule dispenser or by the use of a vibratory tray or corrugated roll feeder. In both cases, only as much of the particulate material is dispensed as needed to coat the cereal pieces on the surface of the product bed.

Formulation

The section in Chapter 10 on the fortification of breakfast cereals gives suggested vitamin blend formulas and spray rates appropriate for Phase I as already described.

For Phase II sugarcoating applications, the formulations are really quite simple and straightforward. The major objective is to combine a sugar formula with an application method that produces a coating of sugar crystals of the desired size and structure when dry and the desired color and flavor. Various ways to accomplish this have been

Figure 21. Powder-granule auger tack-on applicator (Hanson, 1985) in operation. (Courtesy Spray Dynamics, Costa Mesa, CA)

the subject of many patents over the years. Initially, the objective was to apply a hard, transparent glaze or candy coating that involved very little added water but required a very hot (even molten) spray solution (Massmann et al, 1954; Vollink, 1959). Green and Smith (1963) added powdered hard candy or sugar to a tumbling mass of base product and heated it to fuse the sugar in place so that it adhered to the cereal pieces. Fast (1967) avoided such protracted high heat by including glucose, acetic acid, sodium acetate, and invert sugar with the sucrose in the spray solution, continuously heating it and depositing it on the product in a ribbon-type mixing conveyor just as it reached the critical temperature range. The dusting on of a powdered proteinaceous material after coating with heated honey or corn syrup was described by Gilbertson (1978).

McKown and Zietlow (1971) claim to make a crisp, nonstarchy, nonglossy coating by mixing powdered sugar and cereal with a 7.5% gelatin spray and tumbling them together, then drying, separating aggregates, and removing nonadhering sugar. Other patents have been issued describing the use of edible fat as a coating layer (Hawthorn, 1974; Cole, 1977; Furda, 1983) or emulsified in a syrupy spray (Lyall and Johnson, 1976). Furda coated with a mixture of crystalline fructose and high-fructose corn syrup, enrobing the product with heated edible oil and then dusting it with dry powdered sugar.

A desirable frosty appearance is claimed by Edwards (1982) from application of both a powdered crystalline sugar mixture and a layer of a concentrated aqueous solution of dextrose and sucrose, and by Verrico (1987) from the simultaneous atomizing and spraying of a sweetener solution, followed by drying of the sprayed particles with compressed air. Schade et al (1978) describe the spraying of a solution consisting of a high-intensity sweetener and an encapsulating colloid while partially evaporating the solvent so as to deposit the sweetener and carrier as a foam on the base product. Removal of the remainder of the solvent then produces a frosty appearance.

The basic ingredient of most sugarcoatings is still cane or beet sugar (sucrose). Frequently it is preground to between 4X and 10X particle size—preferably without anticaking agents such as cornstarch. Mixtures of as much as seven parts of sugar to one of water can now be pumped and sprayed at only slightly above room temperature.

Small amounts of oil may also be included in the formulation to assist in preventing clump formation. Other flavoring materials can also be added. Brown sugar or honey can replace part of the white sugar for flavor effects. Natural and artificial flavors may be added and, in the case of fruit flavors, citric and/or malic acids included as enhancers.

Conclusion

The role of coating operations in the breakfast cereal industry is an important one. In the United States, about one third of all ready-to-eat cereals are presweetened. Most are also fortified with vitamins and minerals. The information presented in this chapter outlines some of the basic principles of operating several coating application systems.

References

Cole, K. M. 1977. Continuous double coating—Natural cereal. U.S. patent 4,061,790.

Edwards, L. W. 1982. Co-crystallization of dextrose and sucrose on cereal products. U.S. patent 4,338,339.

Fast, R. B. 1967. Process for preparing a coated ready-to-eat cereal product. U.S. patent 3,381,706.

Furda, I. 1983. Method for preparing food products with sweet fructose coating. U.S. patent 4,379,171.

Gilbertson, D. 1978. Sweet coating for food products. U.S. patent 4,089,984.

Green, J., and Smith, P. S. 1963. Sugar-coating process. U.S. patent 3,094,947.

Hanson, H. W., Jr. 1981. Clog-free spray system. U.S. patent 4,283,012.

Hanson, H. W., Jr. 1985. Tack-on applicator. U.S. patent 4,493,442.

Hawthorn, L. 1974. Process for preparing a coated ready-to-eat breakfast cereal. U.S. patent 3,814,822.

Lyall, A. A., and Johnson, R. J. 1976. Emulsified oil and sugar cereal coating and incorporating same. U.S. patent 3,959,498.

Mancuso, James, Jr. 1983. Clog-free spray system. U.S. patent 4,422, 574.

Massmann, W. F., Michael, E. W., and Vollink, W. L. 1954. Process for hard candy coating food particles. U.S. patent 2,689,796.

McKown, W. L., and Zietlow, P. K. 1971. Process for sugar coating ready-to-eat cereal. U.S. patent 3,615,676.

Schade, H. R., Baggerly, P. A., and Woods, D. R. 1978. Frosted coating for sweetened foods. U.S. patent 4,079,151.

Verrico, M. K. 1987. Method and apparatus for spraying snow-like frosting onto food stuff particles. U.S. patent 4,702,925.

Vollink, W. L. 1959. Process of producing a candy coated cereal. U.S. patent 2,868,647.

CHAPTER 8

Package Materials and Packaging of Ready-to-Eat Breakfast Cereals

EDWARD J. MONAHAN
ELWOOD F. CALDWELL

Packaged ready-to-eat cereals were first sold in retail stores around 1910. The original packages, described as "large size" (6–9 oz.), were prefabricated waxed glassine bags. The bags were filled by hand, closed with a double fold and a wax heat seal, and inserted into printed paperboard cartons by hand. Moisture protection was adequate for the anticipated life of the packaged product. The 1-oz. single serving, or restaurant style, followed later. At first the package was a printed, solid, bleached sulfate carton without a liner. The cartons were filled, sealed with cold glue, and then dipped into molten paraffin wax. Again, moisture protection was adequate, but the operational rate was less than 20 cartons per minute. The number of people needed to operate either kind of packaging line was very high because the owners typically did not want to invest in packaging machinery.

In the 1930s a machinery manufacturer first developed and offered equipment that mechanically formed liners from a roll of waxed glassine and plunged them into a carton (with the bottom and side seam glued) at the rate of 30 cartons per minute. The lower package material costs more than compensated for the initial expense of purchasing the equipment. Breakfast cereal companies were the first to install these machines. Modules of machinery soon followed that folded and tucked the liner into the top-opened carton and weighed, filled, and sealed the carton. With experience, operation of these package lines became increasingly efficient.

Multiple Functions of Packaging

Although the package styles of today may not seem much different from their forerunners of more than 70 years ago, the objectives of packaging have multiplied. In addition to *product protection* and *product identification,* major functions of today's package include *consumer attraction* at the point of purchase and *consumer appeal* throughout use in the home.

PRODUCT PROTECTION

Product protection begins with the filling operation at the packing line and ends when the consumer empties the package. The package must protect the contents from breaking and from the entry of moisture and contaminating odors. A recent objective, equally important, is protection from loss of or change in flavor, particularly in aromatically flavored cereals.

The rigidity of the carton stock and the compression resistance of the finished carton in a well-designed shipping container together must prevent product breakage throughout production line operations, warehouse storage, and distribution from the manufacturer to the retailer. Rigidity also stops the carton from bulging, which is important because any increase in carton volume due to bulging results in apparent slack fill and consumer perception of improper product weight.

The lining paper or film must fulfill all other product protection requirements. It must be flexible enough to withstand the packaging line without puncturing but rigid enough to provide seals without wrinkles, which could allow moisture to penetrate, or product crumbs, which can also affect seal integrity. It must resist the penetration of water vapor and the transport of gases or vapors in either direction.

PRODUCT IDENTIFICATION

The product name should stand out from all other parts of the graphic design and grab immediate attention. Because most cereals are "presold" to consumers before they enter the store, through television and newspaper commercial advertising, the name on the package should serve as an attractive reminder. Other aspects of product identification on the package include the name and address of the manufacturer, the net weight statement, and (in the United States and several other countries) the list of ingredients and the nutritional labeling (nutrients per ounce and serving).

CONSUMER ATTRACTION AT THE POINT OF PURCHASE

To attract consumers, part of the package graphics is devoted to portraying how the product will look when served, highlighting any premium offers, or conveying whatever distinctive concept the product provides. This mission must be accomplished in a matter of seconds— the time the shopper stands in front of that carton on the shelf.

The responsibility for the final design is usually shared among the in-house advertising department or other unit, the product manager, and the external advertising agency. What they want typically dictates the best surface for the paperboard and the best method of printing for optimum print quality.

CONSUMER APPEAL THROUGHOUT THE USE OF THE PACKAGE

Consumer satisfaction with the package is largely related to how well the package protects the product, how easily the package opens, and how well the consumer can control the pouring of the product into a serving bowl. How efficiently the liner refolds to protect the opened product and how neatly the carton closes for storage on the kitchen shelf also have an impact on the consumer. All cereal cartons claim to be easy to open and reclose, which requires a high standard of glue application that is not always met by production operations. However, the cartons perform well most of the time, as glue application has become more efficient and reliable with the use of hot-melt adhesives. Although these design features may or may not completely satisfy the consumer, they add little or nothing to the cost of the carton.

Package Components

The materials used to package breakfast cereals typically include printed paperboard cartons, protective liners, corrugated shipping containers, and the necessary adhesives.

CARTONS

Paperboard manufacturers continually strive to develop a stiffer paperboard without increasing the caliper (thickness) or weight and to maintain or improve printing quality without increasing the cost. The typical paperboard stock is white patent-coated newsboard (WPCN). The white side is usually made of two plies of bleached groundwood pulp with some trimmings from No. 1 bleached sulfite envelope stock and office papers. A binder, such as polyvinyl acetate resin, is used to increase the holdout of the inks for better quality and uniformity. Calcium carbonate is added to make the surface smoother for improved printing. Clay is sometimes used on the calender stacks, and the resulting board is then called machine clay-coated newsboard (MCCN).

Cereal cartons were originally printed by the letterpress method, which is no longer used to any extent. The lithographic process gives better reproduction, print quality, and surface smoothness and is the most widely used process today. Rotogravure printing also offers excellent reproduction at lower cost than lithography if the requirement of long runs or very high volume can be met. The addition of a flexographic station on the rotogravure press allows frequent changes in the back panel graphics for new premiums or recipes.

With lithographic printing, the paperboard stock is sheetfed and automatically stacked on pallets after printing. The printed sheets are delivered to the carton area, where they are fed into a cutting and creasing press. The sheets are then stacked on a pallet and stripped of trim with a hand-operated hammer knife (Figure 1).

Stock for rotogravure printing is roll-fed into the printing press and exits into in-line rotary cutting-creasing rolls or reciprocating flatbed cutting and creasing forms (Figures 2 and 3). The cartons are then automatically stripped and stacked on a pallet for shipping. This system saves labor, but both the printing rolls and the cutting and creasing rolls are very expensive, limiting rotogravure to high-volume products for major carton users.

The pallets containing either kind of carton are wrapped in plastic to help maintain stable orientation and to protect the cartons from dirt and moisture changes during shipping. Cartons are delivered on standard pallets in tiers about 4 ft high, which weigh about 1,200 lb.

The cartons delivered to cereal processors are tested against specifications for odor, stiffness, caliper, dimensions, crease embossing, board color and weight, ink color uniformity, and overall printing quality. In some cases, water absorbency is also specified to permit the use of water-based adhesives. Typical specifications are shown in Table 1.

The carton industry caters to the cereal industry. Carton suppliers are frequently (but not always) located near cereal plants to keep freight costs at a minimum for low-cost bidding on purchase orders, as well as to assure proper supply throughout the year in response to changing circumstances. They generally upgrade their operations as soon as improved presses become available. The high volume and relatively steady supply requirements necessitate dynamic decisions by the carton suppliers to do what appeals to the cereal carton buyers.

LINERS

The lining material is the "workhorse" of the package composition; yet it is also the thinnest, the most abused in production operations, and by far the lowest in cost except for adhesives. The lining material must perform on the packaging machines and make good seals without

Figure 1. Sheetfed cartons printed lithographically are stripped of trim by a hand-operated hammer knife after being stacked on a pallet. (Courtesy Waldorf Corp., St. Paul, MN)

Figure 2. Roll-fed carton stock exits from a rotogravure press to in-line cutting and creasing. (Courtesy Waldorf Corp., St. Paul, MN)

TABLE 1
Typical Cereal Carton Board Specifications (Partial) for 20-Point Machine Clay-Coated Newsboard (MCCN)[a]

Attribute	Unit of Measure	Test Procedure[b]	Minimum Value	Maximum Value
Basis weight	lb/1,000 ft^2	TAPPI T410	78.0	88.0
Caliper, cartons	0.001 in.	TAPPI T411	19.0	21.0
Moisture, cartons	%	ASTM D464	5.0	7.0
Taber stiffness[c]				
MD	Taber units	ASTM D781	230	...
CD	Taber units	ASTM D781	60	...

[a]General description: Board stock shall be all from the same mill run within a purchase order, free of objectionable or contaminating odors, able to pass through metal detectors calibrated with 3/32 in. Series 400 stainless steel sphere, and have a news side of clean appearance and a post fiber bond between the top liner with its clay content and the index liner. Material shall be of food grade and in all respects in compliance with the federal Food, Drug and Cosmetic Act of 1938 as amended and all applicable regulations thereunder, including conditions of manufacture, storage, and shipment.
[b]References are to methods published by the Technical Association of the Pulp and Paper Industry (TAPPI) and the American Society for Testing and Materials (ASTM).
[c]MD = machine direction, CD = cross direction.

any openings or wrinkles. It must not permit the entry of moisture or moisture vapor, even after the line abuse. In most cases, it acts as a barrier to trap cereal aromas within the product and at the same time prevent foreign odors from penetrating the packaged product. It should not irritate the consumer by tearing or allowing the product to fall between the carton and the liner. It should also be reclosable to protect the cereal remaining in the package.

In a few cases, where the cereal product is not hygroscopic and/or retains satisfactory texture in moisture equilibrium with the ambient atmosphere, a liner may not be needed for moisture protection and may even entrap rancid aromas. In such cases, either no liner or a vapor-permeable liner may be used. Some shredded wheat products are in this category.

The major lining material used today for cereals is plastic film. High-density polyethylene (HDPE) is the most common because of its

Figure 3. Automatic in-line cutting, creasing, and stripping of rotogravure-printed carton stock. (Courtesy Waldorf Corp., St. Paul, MN)

availability and low cost. HDPE coextruded with a thin film of ethyl vinyl acetate (EVA) is a typical combination. EVA allows a low-temperature seal that does not affect the HDPE, and it offers the consumer an appealing and peelable seal. A partial liner specification is shown in Table 2.

The supercalendered waxed glassine that at one time was the dominant material for cereal package liners has been partially replaced by plastic film today. Some cereal processors may use "in-line" calendered glassine more to accommodate existing package-making machinery than to meet product requirements. In some cases, waxed liners are used by cereal processors who have integrated package material operations and who supply their own waxed materials. A typical waxed glassine specification is shown in Table 3.

The availability and advantages of plastic films were the reasons for the decline of glassine paper systems. Papermakers were forced to modify their Fourdrinier glassine papermaking machines to meet

TABLE 2

Typical Cereal Carton Liner Specifications (Partial) for 2.2-Mil Coextruded High-Density Polyethylene (HDPE) and Ethyl Vinyl Acetate (EVA)[a]

Attribute	Unit of Measure	Test Procedure[b]	Minimum Value	Maximum Value
Caliper	0.001 in.	TAPPI T411	2.1	2.3
Finished weight	lb/3,000 ft^2	TAPPI T410	30.1	35.3
HDPE weight	lb/3,000 ft^2		24.1	28.2
EVA/Surlyn weight	lb/3,000 ft^2		6.0	7.1
Tear strength[c]				
MD	g	TAPPI T414	15.0	...
CD	g	TAPPI T414	350	...
WVTR[d]	g/100 in.^{2}/24 hr	ASTM E96	...	0.28
Coefficient of friction				
Sealant to sealant	...		...	0.60
Poly to poly	...		...	0.20
Heat seal strength	lb/in. width		2.0	...

[a]General description: Liner stock shall be 2.2-mil coextruded 80/20 HDPE/EVA film, food grade and odor-free, and of uniform gauge, as free as possible of pinholes and processed to give acceptable machine operations on Hayssen Vertical and Triangle machines. The finished weight shall be 32.6 lb (HDPE 26.1 lb, EVA/Surlyn 6.5 lb), roll diameter 18 in. maximum, core diameter 3 in. Material shall be in all respects in compliance with the federal Food, Drug and Cosmetic Act of 1938 as amended and all applicable regulations thereunder, including conditions of manufacture, storage, and shipment.
[b]References are to methods published by the Technical Association of the Pulp and Paper Industry (TAPPI) and the American Society for Testing and Materials (ASTM).
[c]MD = machine direction, CD = cross direction.
[d]Water vapor transmission rate.

the new functionality standards for the cereal trade or lose the entire business. To compete with the plastic films, they have resorted to longer fiber lengths (reduced Jordaning), resin additives, and coatings to make the glassine stronger, more flexible, and less moisture-sensitive. In some cases, glassine still outperforms plastic in terms of barrier protection and reclosability. Also, some cereal packaging equipment cannot accommodate the plastic films.

As discussed in Chapter 10, cereal cartons and liners, in addition to extending product shelf life via their inherent barrier properties, also provide a way to incorporate antioxidants into the product-package system. If an antioxidant is used, it must be declared on the label.

TABLE 3
Typical Cereal Carton Liner Specifications (Partial)
for Laminated and Overwaxed Cereal-Grade Glassine[a]

Attribute	Unit of Measure	Test Procedure[b]	Minimum Value	Maximum Value
Finished weight	lb/ream	TAPPI T410	50.3	55.7
Glassine weight	lb/ream	TAPPI T410	18.0	20.0
Wax weight				
Total	lb/ream	ASTM D590	19.0	25.0
Laminate	lb/ream	ASTM D590	13.0	17.0
Overwax	lb/ream	ASTM D2423	6.0	8.0
Tear strength[c]				
MD	g	TAPPI T414	24.0	···
CD	g	TAPPI T414	28.0	···
WVTR[d]				
Flat	g/100 in.^{2}/24 hr	ASTM E96	···	0.10
Fold	g/100 in.^{2}/24 hr	ASTM E96	···	0.30
Tensile strength[c]				
MD	lb/in. width	ASTM 0882	28.0	···
CD	lb/in. width	ASTM 0882	13.0	···
Gurley stiffness[c]				
MD	mg		90.0	···
CD	mg		45.0	···
Overwax melting point	°F	ASTM F766	135	143
Moisture	% of paper weight	ASTM D644	4.5	7.0

[a]General description: Liner stock shall consist of two 19-lb cereal glassine sheets laminated with 7-lb wax and overwaxed with 8-lb wax to a finished weight of 53 lb, processed to give low water vapor transmission rate in flat and fold and acceptable machine operation on a Pneumatic Double Package Maker, with roll diameter 20 in. maximum and core diameter 6 in. Material shall be in all respects in compliance with the federal Food, Drug and Cosmetic Act of 1938 as amended and all applicable regulations thereunder, including conditions of manufacture, storage, and shipment.
[b]References are to methods published by the Technical Association of the Pulp and Paper Industry (TAPPI) and the American Society for Testing and Materials (ASTM).
[c]MD = machine direction, CD = cross direction.
[d]Water vapor transmission rate.

SHIPPING CONTAINERS

Shipping containers are usually made in three sizes: 12-unit, 24-unit, and 36-unit. Thus they can accommodate low-volume items for small stores or stores catering to small households as well as high-volume items in large supermarkets. (Some pallet loads of stacked cartons are produced for special-volume customers but not in significant amounts relative to total industry volume.)

All cereal shipping containers are made with a B or C flute, and all are of the so-called regular slotted container (RSC) construction (Figure 4 and Table 4). They are generally made in the vicinity of the cereal plant because their large volume and light weight make it cost-prohibitive to ship them any distance.

Experimental shipments of corrugated trays and plastic wraps of 12- and 24-carton shipping units have been tried. However, unless the pricing of corrugated stock drops significantly, this experimental style is not likely to be adopted for regular cereal production.

ADHESIVES

The cereal industry is the second largest user of adhesives for consumer products, behind the tobacco industry. Cereal processors used slow-setting, cold-water starch adhesives in the early days of hand operations. Today, water-based emulsions are used in most operations, and hot melts are also used to some extent. The use of hot melts is growing because of their surface adhesion and quick set and the sophistication of the glue guns mounted on the equipment.

Packaging Equipment

The packaging system of the cereal packaging line begins with weighing the product and ends with the delivery of the pallet load to the warehouse. At all points on the production line, careful attention must be paid to protecting the product from breakage and abrasion.

Compared to many food product lines, cereal packaging systems are fairly slow. Most cereal lines run at no more than 40 cartons per minute, but because intensive labor is not required, the cost of the cereal package is still relatively low. (As indicated later, modular bag-in-box lines can be run at up to 80 cartons per minute by adding a second bag former and filler.)

As illustrated in Figure 5, cereal packaging systems produce either lined, printed cartons (lines I and II), printed cartons with a pouch (line III), or printed, flexible pillow pouch bags (line IV).

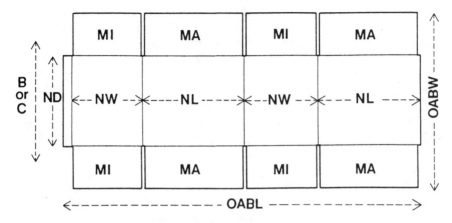

Figure 4. Generalized diagram of a regular slotted container (RSC) style corrugated shipping container before setup (not drawn to scale). OABW and OABL, overall blank width and length; NW, NL, and ND, nominal width, length, and depth; MA, major flaps; MI, minor flaps; B or C, flute type and direction of corrugation. Note that MA = MI in depth, minimizing scrap, while NL > NW, to provide overlap for stacking pattern on pallet.

TABLE 4
Typical Specifications for Regular Slotted Container (RSC)
Corrugated Shipping Container for Cereal Packages[a]

Attribute	Unit of Measure	Test Procedure[b]	Minimum Value	Maximum Value	Target Value
Compression					
Top/bottom	lb	ASTM D642	475	⋯	530
Side/bottom	lb	ASTM D642	725	⋯	805
End/end	lb	ASTM D642	715	⋯	795
Deflection at failure	0.001	ASTM D642	⋯	500	⋯
Edge crush	lb/linear in.	TAPPI T811	32	⋯	36
Flat crush	psi	TAPPI T810	36	⋯	40
Basis weight	lb/1,000 ft^2	TAPPI T410	52	⋯	⋯
Mullen burst test	psi	TAPPI T810	125	⋯	⋯

[a]General description: container style, RSC; additional components, none; container closing system, hot melt; machinery applications, Douglas case packer; nominal length, 13-11/16 in. inner diameter (i.d.); nominal width, 12-7/8 in. i.d., nominal depth, 9-11/16 in. i.d.; overall blank size 22-13/16 × 54-7/8 in.; scoring type (flap), point to point, offset 1/32 in.; scoring type (body), point to flat, flexo score; flap folding sequence, minors in, majors out; manufacturing joint position, outside; manufacturing joint extension, none; flute designation, B; corrugation direction, normal; number of corrugated walls, one; liner type, natural kraft; type of combining adhesive, regular starch.

[b]References are to methods published by the Technical Association of the Pulp and Paper Industry (TAPPI) and the American Society for Testing and Materials (ASTM).

LINED, PRINTED CARTONS

Lined, printed cartons are made on one of two types of packaging lines. With a high-speed bottom sealer and plunger, the liner is formed around a mandrel and then plunged into a carton with bottom and side seals (Figures 6 and 7). With this method of package forming, short runs can be made without costly and time-consuming mechanical modifications. Its operation is less complex than that of a Double Package Maker (Pneumatic Scale Corp., Quincy, MA), and it has been considered the more efficient of the two. Machinery manufacturers have begun to update this equipment because of its efficiency, reliability, and simple maintenance requirements.

With a Double Package Maker, the liner is also formed around a mandrel, but a flat carton is placed above it and wrapped around the same mandrel, and a lined carton is ejected (Figure 8). This method was developed to replace the bottom sealer-plunger method because flat cartons in tied bundles are cheaper than knocked-down flat cartons (KDF) in corrugated containers.

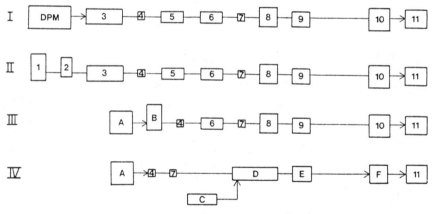

Figure 5. Schematic flows of four cereal packaging systems. Each results in a plastic-wrapped pallet of cased packages ready for shipping. In lines I and II, the package is lined, printed newsboard carton; in line III, it is a printed carton with a pouch liner; in line IV, it is a printed flexible pillow pouch bag. DPM, Double Package Maker; 1, carton bottom sealer; 2, liner former and inserter; 3, product scales and filler; 4, checkweigher; 5, liner sealer and tucker; 6, carton top closer; 7, metal detector; 8, carton collating and case packing; 9, case sealing; 10, automatic pallet loader; 11, pallet plastic wrapper; A, product scales and pouch form-fill-weigh-seal; B, carton erector, pouch inserter, and carton sealer; C, area for manual case setup and cell insertion; D, area for manual pouch insertion; E, case sealer; F, manual pallet loading.

PRINTED CARTON WITH A POUCH

The bag-in-box style (Figures 9 and 10), a revision of an old method, was introduced as a cereal line after the advent of plastic films for the bag or pouch. It has the potential to permit easy mechanical changes in the pouch size over a wide range of volumes and to run at a speed almost double that of other methods when two vertical bag formers and fillers are used. Horizontal cartoners are available that can handle the output from three or more bag formers. These systems have found acceptance among the large cereal processors.

With increasing need for flexibility to meet changing marketing requirements, a parallel development is the use of a slower (40–60 cartons per minute) but more flexible integrated system with a single former-filler, as shown schematically in Figure 11. Several of these units can be used separately or arranged as a line to increase production while maintaining flexibility.

Figure 6. High-speed bottom sealer for cereal cartons. Knocked-down flat (KDF) cartons with preglued side seam enter from magazine (near right) and are opened and bottom-glued, emerging ready for liner (far left). (Courtesy Pneumatic Scale Corporation, Quincy, MA)

PRINTED PILLOW POUCH

The printed pillow pouch is more than 50 years old but has been used for cereals only since the late 1970s, with the introduction of

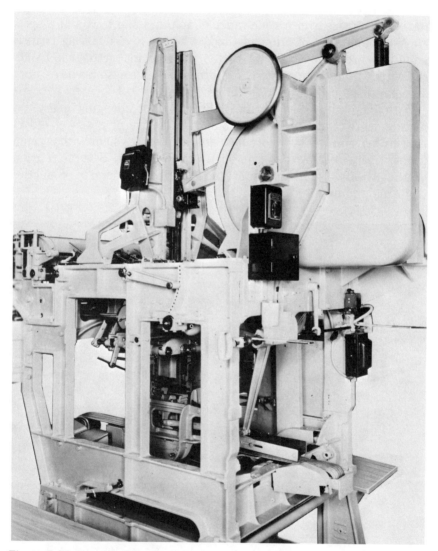

Figure 7. Plunger machine for lining cereal cartons (cartons not shown). Open-top cartons from bottom sealer enter at lower near right. Liner is fed from a roll (center left), cut off, wrapped around a vertically reciprocating mandrel, and plunged downward into carton, which, when released, emerges (lower far left) for filling. (Courtesy Pneumatic Scale Corporation, Quincy, MA)

inexpensive, quality-printed plastic films. A typical form-fill-seal pouch machine is pictured in Figure 12.

The printed pillow pouch is the least expensive style in materials and package machinery costs. However, it is the most labor-intensive because in order to withstand distributor warehouse stacking of pallet loads, the containers must have cell support inserts. It is not yet mechanically feasible to set up the container, put in the inserts, and load the flexible pouches into the individual cells automatically. There are soft-touch carton packers that can load a case of 24 bags automatically without inserts. The cases can then be palletized, but the pallets cannot be stacked. As a result, the pillow pouch method is used by small, regional cereal processors who distribute to stores directly and even reuse the RSC shipping container. The cost advantage in this method of merchandising and distribution within a small region is considerable.

Figure 8. Double Package Maker: liner is fed from roll stock (near right) and formed around a mandrel; a flat carton (from center left) is wrapped around it on the mandrel; the side seam and bottom flaps are glued; and the lined carton is ejected (far right). (Courtesy Pneumatic Scale Corporation, Quincy, MA)

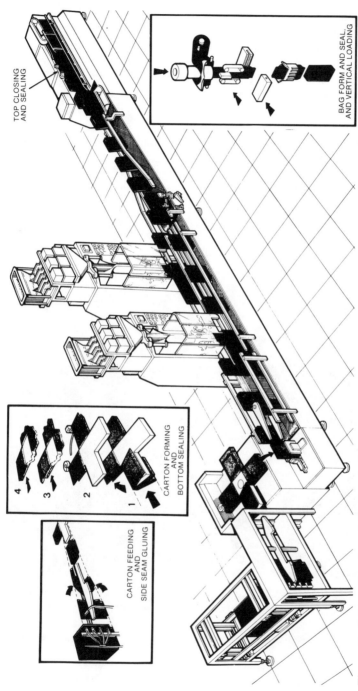

TOP CLOSING
AND SEALING

BAG FORM AND SEAL,
AND VERTICAL LOADING

CARTON FORMING
AND
BOTTOM SEALING

CARTON FEEDING
AND
SIDE SEAM GLUING

Figure 9. Schematic of modular bag-in-box line. Two vertical bag formers and fillers working in parallel insert filled bags into cereal cartons at the rate of 80 or more cartons per minute (each filler at 40 cartons per minute). Flat printed cartons are fed from the magazine at far left through a side-seal glue applicator into a four-mandrel carton former and bottom sealer. Cartons then drop alternately into the two conveyor lines with their open tops facing upward to receive a filled and sealed bag from one of the vertical bag machines (center). The two lines then merge again for top closing and sealing (far right). (Courtesy Pneumatic Scale Corporation, Quincy, MA)

The Cereal Packaging Line

Scales (see Figure 5) ensure proper product weight for the buyer and efficiency for the processor. Electronically operated checkweighers allow small overweights but no underweights. This double-check benefits the consumer and assures the processor of line efficiency and a true measure of product expense.

Liner sealers (Figure 5, lines I and II) close the liner by heat-sealing the inner sides. The upright lip is folded down, and the ends are tucked inside the carton. A few machines still double-fold without sealing, which requires more liner material and is therefore more expensive. Sealers for smaller cartons use a rotary motion for faster operation in less floor space.

The carton top sealer glues with hot melt. The glue applicator is designed to permit easy opening and reclosing of the carton by the consumer. The use of hot melts has eliminated the need for specifying water absorbency of the carton stock, which can be a factor in improved

Figure 10. Part of a modular bag-in-box line in operation. (Courtesy Pneumatic Scale Corporation, Quincy, MA)

printing quality, and has shortened the compression belts on the machine and increased its efficiency.

The case collator takes a shipping case from a magazine and sets it up squarely, with the flaps held open. One or two tiers of cartons are collated and plunged into the case with one stroke. The case flaps are folded and heat-sealed. Because of the efficiency and slow speed of the collator, one person can easily operate two case-pack areas.

Automatic palletizing is performed in the shipping area but is considered part of the packaging line. The cases are received and placed on the pallet automatically according to a preprogrammed pallet design. The design must consider the size and weight of the filled case to allow for maximum resistance to crushing, given the height of the pallet stacking required by the distribution system. The pallet is then stretch-wrapped and moved to the storage area.

In lines that use a pouch as a carton liner or as the basic package (lines III and IV in Figure 5), the vertical pouch is formed from a roll of liner stock around a tube; the vertical back seam is either sealed with an overlap of the running edges or formed as a fin seal. The resulting pouches may be either in pillow form to be used as the basic package (line IV in Figure 5) or side-gusseted to better match a carton cross section. Pouches may be advanced (downward) by belts that move the film down the forming tube to stationary cross-seal jaws or by pulling of the draw bar that incorporates the cross-seal jaws. In either case, the jaws form the top seal of the pouch just filled and the bottom seal of the succeeding pouch, and the pouches are separated by knife cutoff.

The pouch (normally side-gusseted) is then conveyed and transferred to a horizontal cartoner, where it is matched to a set-up carton and gradually inserted into the carton for subsequent top and bottom sealing. In an alternative method, an open-top carton is conveyed directly beneath the pouch former, and pouches are transferred vertically into the carton.

Some vertical pouch machines have been modified to produce a reclosable pouch with a plastic zipper-type closure. Either existing or new equipment may be modified.

Advances in Cereal Packaging

The chemical industry has become entrenched in the packaging industry by developing plastic resins for films that are easy to handle and by promoting systems that allow in-house manufacture of plastic liners and even the in-line manufacture of packages for immediate filling and sealing. The best example of such in-line manufacturing is the

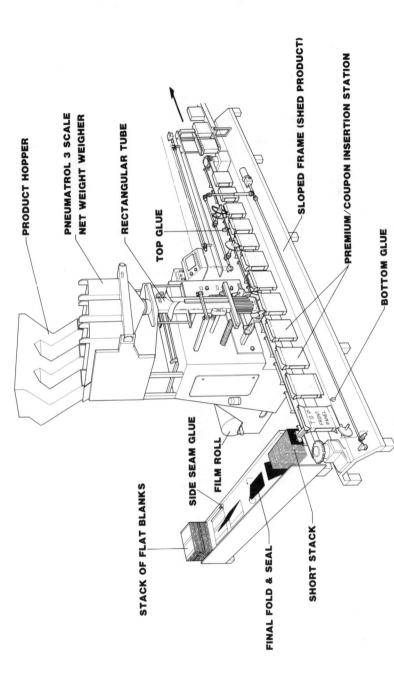

PRODUCT HOPPER

PNEUMATROL 3 SCALE

NET WEIGHT WEIGHER

RECTANGULAR TUBE

TOP GLUE

SLOPED FRAME (SHED PRODUCT)

PREMIUM/COUPON INSERTION STATION

BOTTOM GLUE

STACK OF FLAT BLANKS

SIDE SEAM GLUE

FILM ROLL

FINAL FOLD & SEAL

SHORT STACK

Figure 11. Schematic of a compact, integrated bag-in-box line. Carton setup, filling, and top closing and sealing are all parts of the unit, which operates at 40–50 cartons per minute, depending on net weight. Cartons from a short stack (middle left) come around the near end of the carton conveyor, are bottom-glued en route to the filler (center), and proceed directly to the top closer and sealer (far right). (Courtesy Pneumatic Scale Corporation, Quincy, MA)

packaging of gallons and half-gallons of milk in plastic bottles by the dairy industry. The blow-molded plastic milk bottle costs less than the waxed or plastic-coated sulfite paperboard carton with the gable top and continues to be used although biodegradability has become an environmental issue. The cereal industry will surely continue to

Figure 12. Form-fill-weigh-seal machine for packaging fragile ready-to-eat cereal in a printed pillow pouch. Portion scale system is mounted directly overhead. (Courtesy Triangle Package Machinery Co., Chicago)

analyze consumer preferences and to monitor developments in this area of packaging technology because of the simplicity of package materials storage, the small amount of storage space needed, and automated delivery to the packing or filling line.

In the institutional trade for 1-oz. individual cereal packages, a new vacuum-formed plastic tray with a peelable aluminum and paper lid has been introduced. This style of package meets the specifications for feeding breakfast to school children. However, the glassine- or plastic-lined carton is still the most common style.

Consumers see any cereal package as a printed carton with an inside sealed liner. They see no reason to differentiate among them, but they do notice whether the liner is plastic or paper. Initial complaints about the use of plastics have diminished for cereals, and plastics have been used for more and more products. Little consumer research has been done on cereal packaging, and what is available has come from packaging magazines that have compiled reactions to current packaged products.

Cereal packaging has been the target of customer complaints and dissatisfaction. Consumers have complained about cartons that tear after being opened, top liners that are "welded" too tightly instead of having peelable seals, inability to control pouring, product that falls between the carton and the liner, liners that fail to roll down satisfactorily, and cartons that fail to reclose properly. However, many consumers do not voice their complaints to the company because they feel all packages are the same, and they have more product loyalty than company loyalty. Yet the question arises, "Is anything being done to improve the package?"

Several major changes have occurred in food product packaging. Plastics have made inroads and demonstrated their superiority to paperboard stock packages. The dairy industry has moved toward exclusive use of plastic tubs and bottles. The prevalence of microwave ovens in home kitchens has boosted the acceptance of plastic plates as a package for heat-and-serve items. Plastic bags with zipper-type closings have become preferred over polyethylene-coated paper for food storage and freezer packaging in the home. The plastics industry has methodically moved into first place as the material supplier for today's and tomorrow's packaging.

What is the ultimate for cereal packaging? Easy-opening and -reclosing mechanisms for plastic pouches are needed. An all-plastic, recyclable container should be possible at a total cost comparable to that of paperboard cartons with plastic liners. The experience with plastic milk bottles could be used to introduce a single-wall, blow-molded, rectangular carton with a peelable film inner cover for controlled pouring and a reclosable snap lid. Such developments could emerge from the drawing boards of plastic suppliers and manufacturers of blow-molding machinery.

References

ASTM. 1980. Methods of the American Society for Testing Materials (ASTM). ASTM, Philadelphia, PA.

TAPPI. 1989. Official Test Methods of the Technical Association of the Pulp and Paper Industry (TAPPI). TAPPI, Atlanta, GA.

Chapter 9

Hot Cereals[1]

ELWOOD F. CALDWELL
MYRLAND DAHL
ROBERT B. FAST
SCOTT E. SEIBERT

Growth in sales volume—spurred by technological delopments, by product acceptability and convenience, and by creative marketing and relentless competition—has resulted in domination of the breakfast cereal market by the ready-to-eat or "cold" cereals discussed in seven of the eight preceding chapters of this book. This is particularly true in the traditional breakfast-eating countries, such as the United States, Canada, and the United Kingdom. Many consumers there think of breakfast cereals only in terms of the ready-to-eat varieties, to the exclusion of the cook-up or "hot" cereal products that the term originally implied.

However, the latter still constitute a sizable volume of inexpensive and nutritious products, the per capita consumption of which in the United States began to grow again in the mid-1980s after several decades of decline. Made from one or another of the same five cereal grains as their ready-to-eat descendants, these cereals are the product of an entirely different technology. It typically involves milling in one form or another, many of the details of which are dealt with in other reference books published by the American Association of Cereal Chemists and are beyond the scope of this one. Wheat- and corn-based cook-up cereals are typically by-products or fractions resulting from the milling of these

---

[1] Principal contributors of the section on oat products, including packaging, Elwood F. Caldwell and Scott E. Seibert; on farina-based products, Robert B. Fast and Myrland Dahl; and on other wheat, corn, and rice products, Elwood F. Caldwell.

grains for other purposes. On the other hand, the milling of oats for human consumption is primarily for breakfast cereals.

Accordingly, in this one chapter we deal briefly with all aspects of the manufacture of cook-up or hot cereals and in some detail with the milling of oats for this purpose. Cereals from other grains are dealt with primarily in terms of the use or adaptation of products of their respective milling processes, with reference to other publications for milling details.

Size of Markets and Trends

After the mid-1960s, the hot cereal market in the United States remained relatively flat, with per capita consumption trending down, although growth in population maintained sales at a relatively stable level. Traditionally, hot cereals have done well in times of economic recession, as they are an inexpensive alternative to other forms of traditional breakfast. Table 1 shows the trend in the United States during 1980–1989. By 1989, oat-based products accounted for over 81% of the market by weight, with most of the balance being wheat-based types. Not counting corn grits, other grains made up less than 1% of the 1988–1989 sales. During the period of time covered by the table, there was significant growth in the oat-based instant category, made

TABLE 1
Sales of All Hot Cereals (Excluding Corn Grits) in the United States[a,b]

	1980/ 1981	1981/ 1982	1982/ 1983	1983/ 1984	1984/ 1985	1985/ 1986	1986/ 1987	1987/ 1988	1988/ 1989
Total pounds, millions	358.0	374.5	367.4	365.2	364.6	352.5	347.5	394.5	449.0
Share									
Oat-based, %									
Cook-up	52.6	52.6	51.8	50.0	47.2	48.1	47.5	47.5	47.5
Instant	18.3	18.8	18.8	22.7	26.9	24.9	24.9	27.6	26.4
Other[c]	.7	.7	.9	1.0	1.1	1.2	1.3	2.3	7.3
Total	71.6	72.1	71.5	73.7	75.2	74.2	73.7	77.4	81.2
Wheat-based, %									
Cook-up	24.1	23.9	23.6	21.6	20.8	20.9	21.3	18.1	15.1
Instant	2.7	2.6	3.7	3.6	2.8	3.7	3.8	3.4	2.8
Total	26.8	26.5	27.3	25.2	23.6	24.6	25.1	21.5	17.9
All other, %									
Total	1.6	1.4	1.2	1.1	1.2	1.2	1.2	1.1	0.9
Total, %	100.0	100.0	100.0	100.0	100.0	100.0	100.0	100.0	100.0

[a]During the 52-week period from July through June of the following year.
[b]Source: SAMI-Burke (1989).
[c]Including oat bran.

up of products with a variety of different flavors. Beginning in the fall of 1987, publicity regarding the relationship between the fiber constituents of oat products and the reduction of serum cholesterol generated significant growth in the consumption of oat-based cereals, especially in the oat bran category. However, as this chapter is being written (1989), brans from other grains—notably corn, barley, and rice—are also being promoted as having the same properties.

Rolled Oats and Related Products

The early history of cultivated oats is not clear. For centuries oats were considered to be a weed in barley and wheat fields. Oats took root in northern Europe and particularly in Scotland (Hoffman and Livezey, 1987). In the 18th and 19th centuries, oat mills made their appearance in Scotland, typically incorporating kilns for roasting the grain. After being heat-processed, the oats went through a so-called groat machine, which removed the hulls by the action of a fan (Reynolds, 1970).

Almost all oatmeal available in the United States during the early 19th century was imported from Scotland and Canada and sold primarily in pharmacies (Webster, 1986). The early oatmeal tended to be very floury, as oat groats were ground on millstones, with sifting to remove some but not all of the resulting flour. Oat milling took a great step forward with the invention of a groat-cutting machine by Ehrrichsen in 1877. Ehrrichsen was employed in an oatmeal mill owned by Ferdinand Schumacher, of Akron, Ohio, who was then known as the Oatmeal King and later was one of the founders of The Quaker Oats Company. This groat-cutting device aided in the formation of a granular or steel-cut oatmeal, which produced a cooked cereal with a superior texture, containing little or no fine flour, even without sifting. Subsequently, steel-cut groats were rolled into flakes to form quick-cooking oats similar to the most popular cook-up oat cereal of today (Marquette, 1967).

PROCESSING OF OATS

As shown in Figure 1, the major steps in the traditional processing of oats include receiving the grain, cleaning, drying, hulling, groat processing, steaming, and flaking (Deane and Commers, 1986; Johnson, 1986; Baecker, 1987; Dorn, 1989).

As discussed in more detail in Chapter 1, the U.S. oat crop has been declining in quantity since the 1950s and fell precipitously in the late

1980s, because of government price supports and acreage quotas favoring the production of other grains (Sissons, 1988). Substantial quantities are being imported from Canada and other parts of the world to meet the needs of breakfast cereal processing. At one time, most

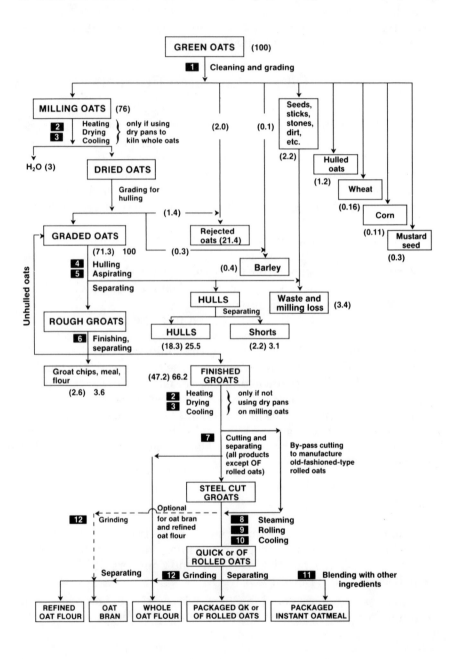

of the crop was used for animal feed. This use resulted in a long-term decline in protein content, with breeders being influenced by feed users and growers being concerned only with quantity (i.e., calories). Because of the milling supply situation and resulting price increases, only half or less of the oats shipped commercially is now used for animal feed. The protein content varies over a range influenced by the country of origin and breeder motivation as well as by crop year, climate, and soil conditions.

Receiving

Oats arrive at the mill by way of bulk railroad car or truck. The first step is sampling to ensure that the oats are of suitable quality for milling. Several samples are taken with a grain probe in an attempt to obtain a representative sample of the oats in the transport vehicle. Among the characteristics typically examined are test weight (lb/bu), sound count, kernel width, and evidence of infestation.

The *test weight* of the oats is very important because it is a measure of the amount of groat or berry present. Milling oats should weigh at least 36 lb/bu and preferably more. Since the groat is the portion processed for human food, a high groat-to-hull ratio is desired. This ratio can vary greatly. Separation of groats and hulls in four samples of oats showed groat percentages ranging from 67.4 to 74.2% (Pomeranz et al, 1979).

The *sound count* is defined by the U.S. Grain Standards (U.S. Code

---

Figure 1. Generalized flow diagram of the milling of oats for human consumption, with numerical data from Barr et al (1955). Figures in parentheses are based on gross yield from green oats as received; figures not in parentheses are based on net yield of dried and graded oats going to hulling. Numbers in boxes show stages: 1 = Cleaning and grading according to size, shape, and density; removal of other grains, foreign material, and hulled oats. 2 = Kiln-drying in a stack of steam-jacketed open pans (to develop flavor; formerly needed to prepare for stone hulling) or heating by direct and indirect steam in a horizontal rotary dryer or vertical grain conditioner. 3 = Cooling before hulling, steel cutting, or rolling. 4 = Hulling in rotary impact huller, formerly between stone discs. 5 = Aspiration, to separate hulls from groats and unhulled oats. 6 = Gravity table, to separate groats from unhulled oats. 7 = Steel cutting, in rotary granulator. 8 = Steaming, by direct or sparging steam at atmospheric pressure, to inactivate enzymes and soften whole or steel-cut groats for rolling. 9 = Rolling between smooth, equal-speed rolls 18 in. in diameter. 10 = Aspiration, to cool and remove fines. 11 = Blending, for instant varieties only. 12 = Grinding and sieving, to make whole oat flour or oat bran and refined oat flour. OF = old-fashioned, QK = quick.

of Federal Regulations, 1988) as including "kernels and pieces of oats kernels (except wild oats) which are not badly ground damaged, badly

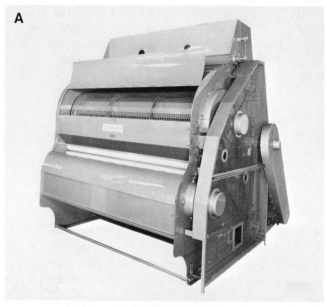

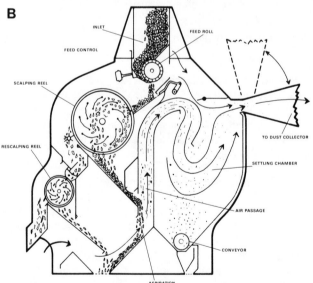

Figure 2. Receiving separator for grain. A, separator with inspection cover open; B, schematic cross-section showing how it works. (Courtesy Carter-Day Co., Minneapolis, MN)

weather damaged, diseased, frost damaged, heat damaged, insect bored, mold damaged, sprout damaged, or otherwise materially damaged."

The *width* of the oat is important because thin oats have a lower groat content. The width is measured with screens with slotted openings.

If evidence of *insect infestation* is found, treatment utilizing approved fumigants must be initiated.

Acceptable oats are then passed over a receiving separator to remove field trash, such as corncobs or other coarse materials, and to remove some of the very fine material, such as chaff dust. One design for a receiving separator is shown in Figure 2. The oats are binned according to milling criteria such as test weight. Data on each bin are maintained so that blends sent to the mill are uniform and give a satisfactory yield and a good-quality end product.

Cleaning and Drying

The objective of the cleaning operation is to send to the mill clean

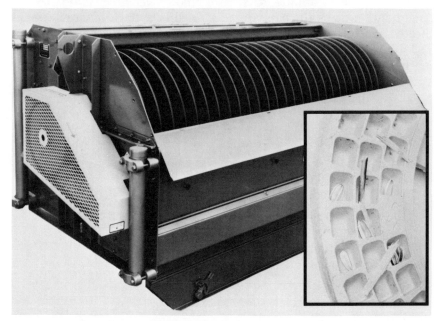

Figure 3. Disk separator with cover removed. This machine separates primarily by length. Disks with indented pockets rotate in a bed of grain and pull out the shorter kernels, leaving the longer ones. Discs are available for a variety of separations. Inset photo is enlarged view of part of a disc, showing how short kernels are retained in pockets on the up side of rotation, to be dumped out on the down side. (Courtesy Carter-Day Co., Minneapolis, MN)

milling oats free of foreign material such as dust, stems, and weed seeds and free of oats that are not satisfactory for milling. Examples of the last category are doubled, pin, light, and hulled oats. Doubled oats contain two groats, neither of which is well developed, and they have a high percentage of hull. Pin oats contain very thin groats. Light oats may be the size of good milling oats but contain a high hull percentage and yield very small groats. Hulled oats are usually weathered and subject to shattering in the hulling process, because they lack the protective hull cover.

The cleaning process utilizes several devices, including screens, aspirators, discs with indent pockets, indent cylinder, and gravity table. These devices take advantage of physical properties of the grain such as overall size (screens), density (aspiration and gravity table), and length or shape (disc and/or indent cylinder machines). The photograph in Figure 3 is of a disc separator with the cover removed; the inset shows a section of one disc.

After completing their trip through the cleaning house, the oats are designated clean milling oats, also known as green oats. In the traditional process the oats are next dried in a stack of circular pans heated indirectly by steam to a surface temperature of 200–212°F (93–100°C). The oats start in the top pan and are moved across its surface by a large rotary sweep. When the oats have completed the

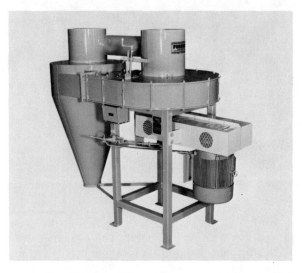

Figure 4. Impact huller. This machine removes groats from hulls by tangential impact of grains against cylindrical outer ring caused by high-speed central impeller. (Courtesy Forsbergs, Inc., Thief River Falls, MN)

trip around the top pan, they drop through a chute to the next pan, and the process continues until they reach the bottom pan. The oats are then binned.

Hulling

The heating process was originally developed to shrink the groat away from the hull and render the hull more friable in order to facilitate hulling with rotating millstones. The set of stones consisted of one stationary bed stone with a millstone rotating above it. The objective was to set the millstones just far enough apart that the oats would tumble end over end and yet the groat would not be ground.

Today, the impact huller (Figure 4) has supplanted stone hulling. The oats are fed through a rotating disc and flung out to strike the wall of the cylindrical housing tangentially, with the hull being

Figure 5. Channel aspirator. Hulls are separated from groats and unhulled oats on the basis of the ratio of surface to weight. (Courtesy Buhler, Inc., Minneapolis, MN)

separated from the oat by the impact. The cylinder wall liner is designed so that the friction between it and the hull approaches or exceeds that between the hull and the groat, causing the latter to separate. Predrying or kilning of the oats is not necessary for this to work.

The mixed material falls to the bottom of the huller and is then subjected to aspiration (Figure 5) to separate the hulls and the hulled groats. Oats not hulled are separated in an indent cylinder or a table separator and sent back for another pass through the huller. The table separator (Figure 6) depends for its effectiveness on differences in specific gravity, surface smoothness, and grain shape. A disc separator (identical to the one in Figure 3 but with different discs) may also be used.

Although the traditional dry-pan process is not a necessity today

Figure 6. Table separator. This machine separates groats from hulled oats by differences in specific gravity, surface smoothness, and grain shape, causing smoother, heavier groats to travel down the table as it oscillates, while lighter, rougher unhulled grains travel upward. Grains are sorted into a series of compartments with zigzag cleats. (Courtesy Buhler, Inc., Minneapolis, MN)

with the impact huller, it is still used by some oat millers, because it may impart a more nutty and less raw or green flavor to the resulting finished products. However, because of the inefficiency of drying hulls (which are then discarded) by a process that is also inherently energy-inefficient, the stack of dry pans has been replaced in most mills by an enclosed vertical or horizontal grain conditioner with both direct (sparging) steam heat (Figure 7) and indirect steam heat (Figure 8). This can be used on unhulled oats but is more typically used to process groats following their separation from the hulls. It imparts flavor to the oats or groats comparable to that resulting from the dry pans.

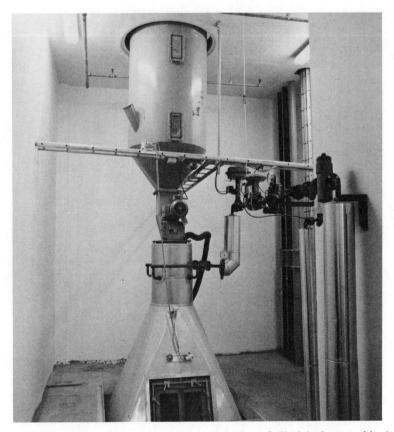

Figure 7. Equipment for conditioning and drying. Cylindrical surge bin (top) feeds into a cylindrical direct steamer (middle) above the heating section of the kiln dryer (bottom), which extends through the floor. (Courtesy of Buhler, Inc., Minneapolis, MN)

Either process effects some reduction in the lipase enzyme content of the groat but not enough to allow further milling without thorough steaming. Oats contain a potent lipase or lipid esterase capable of hydrolyzing not only oat lipids but also those of other ingredients with which rolled oats or oat flour might subsequently be mixed, if it is not inactivated before grinding or rolling.

Groat Processing

After a stream of kiln-dried, dehulled groats has been produced, the next step is to size the groats (Figure 9). Before being sized, the groats

Figure 8. Heating section of vertical kiln dryer (of which the top portion is shown in Figure 7). After being heated by indirect steam, the kilned grain continues to move downward through a cooling section before being binned or conveyed to groat cutters or flaking rolls. (Courtesy Buhler, Inc., Minneapolis, MN)

may be polished in a horizontal cylinder with a center-rotating shaft equipped with paddles (Figure 10). These serve to move the groats through the cylinder, which is lined with an abrasive material. This process removes any remaining hull. Particular mills have specific requirements, but generally the largest, plumpest groats are used for the so-called old-fashioned oats, which are rolled from whole groats. The remaining groats are steel-cut. This process utilizes revolving drums with countersunk holes through which the groats fall to be cut by stationary knives positioned on the outer surface of the drum (Figure 11). It is a descendant of the groat-cutting machine that represented the first great step forward in oat milling in the late 19th century. Typically, a groat can be cut in three to five pieces, with a wide distribution of particle sizes resulting. The particle size distribution is influenced by the size of the groat and the speed of the surface of the revolving drum.

Steel-cut oats cook considerably faster than whole groats and are sold today in specialty shops. They take up to 30 min to cook in boiling water to a porridge that has a distinctive chewy texture, with the steel-cut groat particle noticeable in the viscous mass of the porridge.

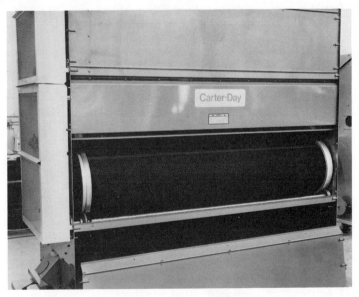

Figure 9. Sizing cylinder, with cover removed. This machine separates groats from other grains and classifies groats by width for cutting and rolling. (Courtesy Carter-Day Co., Minneapolis, MN)

Figure 10. Pearler-whitener. Any remaining hull and fuzz are removed from groats by mild abrasion. (Courtesy Buhler, Inc., Minneapolis, MN)

Figure 11. Groat cutter. Each groat is cut into three to five pieces by stationary knives mounted against the surface of revolving drums with countersunk holes, into which the groats fall. These steel-cut groats are then steamed and rolled to make quick-cooking rolled oats. (Courtesy Buhler, Inc., Minneapolis, MN)

Rolled Oat Flake Production

The production of old-fashioned oats and quick oats is similar. The major difference is the starting material, with whole groats being used for the old-fashioned oats and steel-cut groats for the quick oats. Both products result from rolling between two cast-iron equal-speed rolls, typically 18–20 in. (45–50 mm) in diameter, in rigid end frames.

To get the feed material in condition to roll, the materials typically pass through an atmospheric steamer above the rolls (Figure 12).

Figure 12. Flaking rolls, mounted under steamer bin and roll-feeder. Over the steamer bin (on the next floor and not shown) is a cylindrical direct-sparging steamer similar to the one in Figure 7 (middle). (Courtesy Buhler, Inc., Minneapolis, MN)

Sufficient retention time must be allowed for the feed stock to equilibrate in temperature and moisture as a result of the injection of live steam. A moisture increase from 8–10% to 10–12% is sufficient to provide satisfactory flakes when the whole or steel-cut groats are rolled. The increased temperature has a plasticizing effect and ensures that lipolytic enzymes are inactivated for product stability. As already mentioned, prior heat conditioning of whole or steel-cut groats can also be accomplished through the use of a groat conditioner or kilns.

The rolls are carefully adjusted to give the flake thickness required by the particular product. Quick-oat products are rolled thinner than old-fashioned oats. Feeder rolls uniformly distribute the whole or steel-cut groats across the roll gap to ensure as uniform a thickness as possible.

Following rolling, the flakes are cooled. A variety of devices can be used; generally, ambient air is drawn through the mass of the flakes

Figure 13. Fluidized-bed dryer-cooler (center right). This machine dries and cools rolled oat flakes before packaging. Drying air comes from fan blower at center left. Cooling air comes from the fan below. Air is exhausted out the large round duct over the fluidized bed. (Photo courtesy of Buhler, Inc., Minneapolis, MN)

(Figure 13). Following cooling, the finished flakes can be directed to packaging bins for holding until they are placed in the appropriate package. Data on the analytical and physical characteristics of some typical milled oats products are given in Tables 2 and 3.

OAT FLOUR

Traditionally, oat flour has been nothing more nor less than ground heat-treated oat groats or rolled oats, both being whole-grain products having almost identical chemical and nutritional composition and varying only in particle size and shape. No separation of any of the groat components takes place. Such flour is typically ground on a hammer mill, but it might be ground between differential-speed smooth rolls, depending on the equipment available. It is typically a fine, off-white product that is somewhat coarser than wheat flour. It is used as an ingredient in other food products, such as ready-to-eat cereals or infant foods, and in some nonfood products (Caldwell and Pomeranz, 1973).

TABLE 2
Typical Analyses of Groats and Some Milled Oat Products[a]

Constituent	In Groats and Whole-Groat Milled Products	In Oat Bran	In Refined Oat Flour (Partially Debranned)	Test Method[b] (Where Applicable)
Moisture, %	9.5	9.5	10.0	44-15
Protein, % (N $\times$ 6.25), dry basis	17.0	20.0	13.0	46-10
Ash, %	1.7	3.0	0.6	08-01
Fat, %	6.5	6.5	6.0	30-20
Crude fiber, %	1.3	2.8	0.5	32-15
Total dietary fiber, % dry basis	11.2	18.0	3.0	32-05[c]
Carbohydrate, %	65.0			By difference
Enzyme test	Negative	Negative	Negative	Qualitative[d]
Thiamin, mg/100 g	0.7			
Riboflavin, mg/100 g	0.11			
Niacin, mg/100 g	0.90			
Vitamin E, IU	2.8			
Vitamin B-6, mg/100 g	0.17			
Folic acid, mg/100 g	0.025			
Calories per 100 g	380			

[a]Data from National Oats Co., Cedar Rapids, IA.
[b]Numerical entries are method numbers from AACC Approved Methods (AACC, 1983).
[c]Modified to be consistent with current AOAC method.
[d]Tyrosinase or peroxidase test (see Chapter 10).

However, the advent of oat bran as both an ingredient in commercial cereal and baked products and a consumer cereal in its own right changed this picture. The separation of an oat bran fraction from milled groats necessarily leaves a partially debranned flour that is *not* a whole-grain product. This has been given different names by different manufacturers, such as high-purity oat flour, refined oat flour, debranned oat flour, or simply oat flour. In view of the lack of a legal definition of such terms, oat flour buyers desiring the traditional whole-grain product should specify exactly that, as well as agreeing upon such analytical specifications as levels of moisture, ash, protein, fat, and total dietary fiber and an appropriate enzyme test to ensure that the heat treatment was adequate.

OAT BRAN

Because oat milling traditionally did not include any separation except that of hulls, no product such as oat bran was available as other than a laboratory curiosity until the 1980s. The morphology of the groat,

TABLE 3
Physical Characteristics of Some Typical Milled Oat Products

Characteristic and Unit	Steel-Cut Groats	Rolled Oats (from Whole Groats)	Quick Oats (from Steel-Cut Groats)	Baby Oats	Oat Flour (from Whole Groats)	Oat Bran
Particle size, %						
Over U.S. 4		10–30	0.5–1.5			
Over U.S. 7		60–90				
Over U.S. 8	10–25		65–75			
Over U.S. 10	40–60	<10	8–16			
Through U.S. 10		<10				
Over U.S. 12	15–30		4–8			
Through U.S. 12			6–12			
Over U.S. 14				>75		<10
Over U.S. 16	2–15					
Through U.S. 16	<5					
Over U.S. 20					0	10–45
Over U.S. 30						30–50
Over U.S. 35					<10	
Through U.S. 35					>90	
Over U.S. 40				5–20		10–40
Through U.S. 40				>5		>5
Flake thickness, 0.001 in.		24–29	15–19	10–14		
Density, lb/ft³		26		18	32	
Hulls per ounce		<5	<5	<5		

[a]Data from National Oats Co., Cedar Rapids, IA.

unlike that of the wheat kernel, provides no basis for the mechanical separation of a distinct bran layer or layers, nor do the protein and fat of the outer layers differ markedly from that of the inner layers or the endosperm or germ, as they do in wheat. Wheat germ, for example, can be separated relatively easily from shorts or feed middlings because of the plasticity resulting from its much higher fat content.

Although a sharp separation remains unattainable commercially, the evidence is that a fraction of ground oats rich in oat bran can be separated by a careful combination of grinding and screening and possibly aspiration (Wood et al, 1989a). Apart from visual inspection or proximate analysis, both of which are unreliable indicators of bran content, the degree of concentration is determined by the relative increase in one or more of such known or assumed oat bran components as total dietary fiber, soluble fiber, and β-glucans. An increase in protein content also occurs (Seibert, 1987).

The reason for the commercial interest in oat bran comes from the published results of clinical studies showing that ingestion of an appreciable amount of it on a daily basis as part of a low-fat diet results in a lowering of serum cholesterol over and above any lowering due to the basic diet (Gordon, 1989; Wood et al, 1989b). Serum cholesterol beyond a certain amount is regarded as a significant risk factor for coronary heart disease.

Just as with oat flour, there is at this writing (1989) no legal or even widely accepted practical definition of oat bran. However, a technical committee of the American Association of Cereal Chemists has recommended the following definition:

> Oat bran is the food that is produced by grinding clean oat groats or rolled oats and separating the resulting oat flour by sieving, bolting, and/or other suitable means into fractions such that the oat bran fraction is not more than 50% of the starting material, and has a total β-glucan content of at least 5.5% (dry weight basis) and a total dietary fiber content of at least 16.0% (dry weight basis), and such that at least one third of the total dietary fiber is soluble fiber.

Limitation of the oat bran fraction to not more than 50% of the starting material refers to extraction rate, in milling terminology. This is difficult to verify without access to milling records, but when accompanied by verifiable analytical data, it is a key element of definition as distinct from product specifications or standards, in view of a known spread in analysis for fiber, β-glucan, and protein among different oat varieties. Unpublished evidence indicates that the extraction rate of commercial oat brans shown clinically to reduce cholesterol was about 40%—40 lb of oat bran obtained from 100 lb of groats (leaving 60% debranned oat flour). Hence the consideration of a 50% maximum. In view of the premium price potentially commanded by oat bran relative to whole groats in any milled form, the applicable extraction rate is of great importance to both processors and consumers.

INSTANT OATMEAL

The thinner flakes resulting from rolling steel-cut groats have for many years provided a very quick product requiring boiling for only 1 min, and by the 1950s this was the dominant consumer oatmeal product. (The old-fashioned or whole-groat flakes are still preferred by some consumers and also form a major ingredient of granola cereals, whether commercial or homemade).

Nonetheless, it was clear that however convenient the 1-min quick product was for consumers, it was less convenient than no cooking at all, as with ready-to-eat products. A major effort to close this gap resulted in the development and introduction of instant oatmeal in single-serving envelopes packed eight or 10 to a carton. The major elements of instancy were five in number—1) still thinner rolled oat flakes, 2) some additional heat treatment of the flakes, 3) premeasured servings (to which premeasured very hot water is added by the consumer), 4) premeasured salt (and, later, other additives), and, most important, 5) an added hydrocolloid gum.

Each of these elements contributed to the quality of the product as prepared by the consumer, the most obvious being thinner flakes and added heat treatment. Premeasured servings and salt (requiring individual serving pouches) made possible preparation in the bowl. The patented addition of hydrocolloid gum (Huffman and Moore, 1961) brought about rapid hydration even with very hot water, to which the cold-water-soluble natural oat gums would not otherwise respond rapidly.

Package directions typically call for the consumer to add one-half cup of boiling water to the contents of one individual serving pouch in a bowl, then stir, and eat after adding milk and, if desired, sugar. Since some cooking and setting of structure occur in the bowl, the flavor and texture are improved if the mixed product is allowed to stand a moment or two, although package directions omit reference to this.

The market success of instant oatmeal products (Table 1) has been followed by a proliferation of varieties containing dehydrated fruit bits or other added flavoring materials. The technology also makes it easy to add vitamin and mineral fortification with little or no risk of ingredient separation.

Packaging of Hot Cereals

Like many foods in the last half of the 19th century, oatmeal was sold in bulk, but it was also one of the first foods to be packaged. As Calkins (1974) pointed out, "The package benefited all parties. The manufacturer was able to identify his goods and thwart substitution; the grocer saved time in weighing and wrapping and his shelves were neater; and the consumer received her food fresh and free from germs."

For many years the spirally wound two-ply fiber tube with a paper label was the standard package for oatmeal in the United States (Figures 14 and 15). The top and bottom of the package consisted of flanged caps, with the bottom one being glued to the tube and the top one held on by a glue bond between the label and the cap. An opening device was fabricated from either string or tape, so that the label was torn to release the top cap. However, these are being displaced by new, easier-opening plastic top caps that have a plastic rim and include a reusable snap-on cap.

Most of the other types of hot cereal intended for stove-top cooking, as well as rolled oats in parts of the world other than the United States, are packed in folding cartons. Traditionally, printed paper wrappers

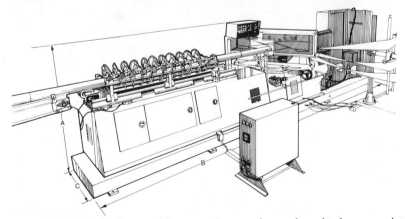

Figure 14. Spiral winding machine to make round paperboard tubes or canisters for rolled oats, with automatic cut-off. (Courtesy Tools and Machinery Builders, Inc., Ironton, MO)

have been applied over blank cartons after filling and sealing to prevent sifting of the contents of the package and chances for infestation. These overwrapped packages perform very well and are still in extensive use. However, a number of manufacturers have converted to printed folding cartons and have achieved sift-proofness and infestation resistance by special gluing patterns on the top and bottom flaps and/or by other special construction features of the package (Figure 16).

Most of the instant hot cereals are packed in individual, single-serving pouches; "measure-your-own" products in bulk cartons are also available. The pouch products are produced on horizontal form-fill-seal equipment or single-lane continuous-motion equipment (Figure 17). Regular products that do not have any added flavors or fruit pieces

Figure 15. Volumetric filling machine for round packages of rolled oats. Paperboard canisters with open tops enter at far left. Product is received into the revolving drum, which elevates it onto a vibrating tray that feeds it gradually into the canisters. During the filling, the canisters are tilted and shaken from side to side to promote settling. The excess rolled oats fall back into the drum and resume the cycle, which minimizes spillage and waste. Filled packages emerge toward the near right and pass over a checkweigher. Adjustment of fill is by changing the shaking speed of the open canisters inside the drum. (Courtesy Solbern Divison of Howden Food Equipment, Inc., Fairfield, NJ)

are packed in a relatively simple structure of paper extrusion-coated with a layer of polyethylene, which forms a seal when heated.

For products containing moisture-sensitive ingredients such as fruit pieces or needing protection from flavor loss, a layer of polyvinylidene chloride is used to increase the moisture barrier and odor barrier properties. This layer is extrusion-coated onto paper before addition of the polyethylene.

Farina-Based Cereals

Second to oat-based hot cereals, the most dominant products are made from farina, defined in the U.S. Code of Federal Regulations (1988) as follows:

(a) Farina is the food prepared by grinding and bolting cleaned wheat, other than durum wheat and red durum wheat, to such fineness that, when tested by the method prescribed in paragraph (b)(2) of this section, it passes through a No. 20 sieve, but not more than 3 percent passes

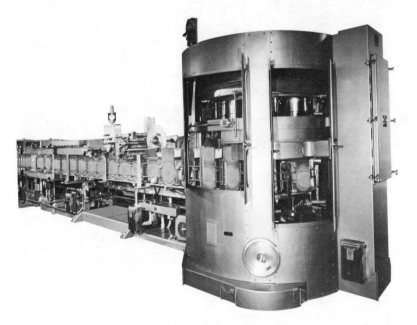

Figure 16. Packager with weight control. This modern automatic combination handles feeding, forming, bottom-sealing, filling, and top-closing of rectangular cartons in one continuous motion to produce up to 400 sift-proof filled cartons per minute. (Courtesy Pneumatic Scale Corp., Quincy, MA)

through a No. 100 sieve. It is freed from bran coat, or bran coat and germ, to such extent that the percent of ash therein, calculated to a moisture-free basis, is not more than 0.6 percent. Its moisture content is not more than 15 percent.

Farina is thus a federally standardized product. For a food product to contain an ingredient identified as farina, the product used must conform to this standard of identity.

The standard goes on in 137.305 to give a standard of identity for "enriched farina." As detailed in Chapter 10, it prescribes the micronutrients that must be added to farina to qualify for that name and the limits of addition for each. No farina labeled as such may be sold that does not conform to one or the other of these standards of identity. In a landmark case in 1943, the Quaker Oats Company unsuccessfully defended its marketing of a product labeled "Farina with Vitamin D," which had been seized by the U.S. Food and Drug Administration. It was neither farina nor enriched farina by legal standards, because

Figure 17. High-speed constant-motion pouch former, filler, and sealer unit (right) feeding through transfer unit (center) to cartoner (left). The machine handles 200–500 pouches per minute. The operational flow diagram (inset) illustrates the continuous nature of the pouch web, which can be of a variety of heat-sealable laminates, depending on product protection requirements. The system is used for packaging instant hot cereal in single-serving pouches in printed folding cartons. (Courtesy R. A. Jones & Co., Inc., Cincinnati, OH)

it contained an enrichment-type ingredient but not the ones required in enriched farina.

In practical terms it can be seen that farina hot cereals are essentially wheat endosperm in granular form. The preferred wheats for farina are hard red spring or winter wheats, because the granules of endosperm stay intact when hot cereals are prepared at home by cooking.

Farina is obtained by drawing off certain streams of product during the milling of those wheats into bread flour, before they are reduced from chunks of endosperm to flour-size particles. In flour milling terminology, they are the first middlings stream. Flour millers prefer not to draw off more than 3% first and 1% second middlings to be used as farina, calculated as percentages of total flour. Since these middlings are prime or high-grade flour stock when reduced in size to flour, removing more than 4% may reduce the baking quality as well as the quantity of the finished bread flour made by the mill.

Middlings from soft wheat flour have also been tried for farina. Because the endosperm of soft wheat is more chalky and less hard and vitreous than that of hard wheats, the granules disintegrate when the cereal is cooked, and the cooked product has a more pasty consistency than cereal made from hard wheat farina.

A typical specification for farina purchased from a mill for blending, processing, or packaging as a breakfast cereal might include the following: moisture, a maximum of 14.0%; ash, $0.35 \pm 0.05\%$; protein ($N \times 5.7$), a minimum of 9.0% (ash and protein being on a 14.0% moisture basis); granulation on U.S. No. 20, a maximum of 1.0%; on U.S. No. 30, a minimum of 50.0%; through U.S. No. 40, a maximum of 7.0%; and through U.S. No. 100, a maximum of 2.0%.

An early producer of farina hot cereals in the United States was the Cream of Wheat Corporation (Minneapolis, MN), whose brand name is sometimes (incorrectly) applied by consumers to all farina, including that made by other manufacturers. Development efforts in the 1930s and 1940s concentrated mainly on the reduction of the cooking time needed to prepare the cereal at home and on nutrient fortification. These efforts led to the patented use of disodium phosphate to reduce the cooking time (Billings, 1938). The enzyme pepsin was also used later for further reduction in cooking. A fortifying process for the addition of calcium and iron was also developed and patented (Billings, 1941). Thiamin, riboflavin, and niacin were added later, along with wheat germ for further nutrient enrichment. This early work was the basis of the present U.S. enrichment standards.

A mild heat treatment (e.g., 140°F, or 60°C, for 15 min) to sterilize the product and head off insect infestation is usually imposed before packaging the product while still hot. *Sterilizing* in this context is an industry term used with reference to insect rather than bacterial contamination. Some flavor development and acceleration of cook-up time may occur as a result of the heat treatment; industry opinions vary as to the existence or extent of this. The equipment used for it is usually a rotary steam-tube heater.

Since the 1960s, further advances have been made that affect farina-based cereals. One was the introduction of mix-in-the-bowl farina cereals, similar to the instant oatmeal products described earlier in this chapter. Various spice and fruit flavors also became possible and were marketed.

The early instant farinas were more than premeasured farina with added salt, since the farina was wetted, pressure-cooked, and then flaked and redried before being packaged in individual portions with added sugar, salt, and dextrin. Proprietary improvements on the basic process continued to be made, to optimize such process conditions as temperatures, moisture content, and cooking time; provide faster hydration; and give improved texture, more like that of the original cooked product. Potential technology for making instant farina cereals was disclosed by Karwowski (1985, 1986, 1987), dealing with process conditions for wetting, tempering, cooking, sizing, flaking, and drying of middlings/farina and with the products that are said to result.

A major improvement related to farina milling came in the late 1970s. Wolffing and co-workers (1979) developed and patented a process for obtaining as much as 25–30% coarse farina from a hard wheat flour mill. Double-tempered wheat is passed through four identical sets of break rolls having deep and coarse corrugations run sharp-to-sharp, in milling terminology. The chop from each break is separated by sifting as in flour manufacture, with substantially oversized pieces sent to the next break as usual. However, particles only slightly oversized are sent to special corrugated rolls for sizing and bran removal before going to the purifiers, from each of which coarse farina can then be drawn in larger than usual amounts.

In addition to forming the basis of portion-pack instant products, farina is also the basic ingredient in other flavored wheat cereals designed for regular stove-top cooking. Its bland flavor and ready availability from a variety of wheat-milling sources makes it ideal for the addition of malt flour, dehydrated malt extract, or cocoa powder. Sweetening and other flavor-blending ingredients may also be included.

Other Hot Cereals from Wheat, Rice, and Corn

As indicated in Table 1 of this chapter, oat- and wheat-based hot breakfast cereals constitute the bulk of the U.S. market, with those from oats over 81% and those from wheat at about 18%, leaving less than 1% for all others (excluding corn grits). Since the majority of wheat products are based on farina, discussed in the previous section, other wheat cereals account for only a few percentage points of the market. This short section deals with those and with other minor products and comments on the manufacture of corn grits and meal and their role as breakfast cereals.

WHOLE WHEAT CEREALS

All aspects of wheat and wheat processing have been thoroughly reviewed recently by Pomeranz (1988, 1989).

Milled Products

Tasty and quick-cooking (but not instant) cereals with a composition similar to that of whole wheat can be made in a hard wheat flour mill by drawing off medium-grind streams such as first- or second-break chop or a medium purifier cut from one of them or a combination. The exact choice depends on the particular mill flow. The right combination makes a tasty, whole-wheat-type cooked product after boiling a minute or two in lightly salted water.

Rolled Wheat

Wheat can be put through a process resembling that used for oat groats, including cleaning, sizing, steaming, and rolling, resulting in a flaked product that resembles rolled oats in the dry form but makes a distinctive cooked product.

Cracked Wheat

Wheat can also be steel-cut or otherwise coarsely ground to make cracked wheat products that require more than a few minutes of cooking but have an attractive nutty flavor and chewy consistency that is less gelatinous than that of cooked steel-cut oats. Mild toasting in a rotary oven or dryer contributes to a distinctive flavor and ensures the destruction of any insect life that might have entered with the wheat. Other grains may also be incorporated.

RICE PRODUCTS

While milled head (whole kernel) rice is certainly a cereal in itself, its place in the diet is that of either a staple carbohydrate food comparable to bread (in the rice-eating countries of Asia), a vegetable to be served with meat, comparable to potatoes (in the Western bread-eating countries), or a pudding ingredient. Its dietary role as a breakfast cereal is thus constrained by culture. However, head rice can be ground into particles about the size of those in farina, which cook up into a specialty hot cereal resembling farina.

With growing dietary interest in fiber, processes for purifying and stabilizing edible rice bran have been developed (Babcock, 1987), and it is available to be used as a fiber-rich food ingredient in other cereals and baked products. It has not yet found independent application as a hot cereal or hot cereal ingredient.

All aspects of rice and rice processing have been thoroughly reviewed by Juliano (1985), with special emphasis on its international aspects.

CORN PRODUCTS

The dry milling of corn resembles that of wheat and rye in that it is a gradual reduction process employing rolls, sifters, and purifiers. However, the objective is quite different: to produce the maximum quantity of coarse grits or meal as the first-choice product and the least possible amount of fine flour. The prices received for these products reflect their relative desirability by consumers and the trade. It is as if a hard spring or winter wheat flour mill were run to produce the maximum amount of first middlings, with flour as a lower-value by-product. Naturally, therefore, corn mill flows contain no smooth rolls, whose sole function is the size reduction that the corn miller tries to avoid.

All aspects of corn and corn milling have been reviewed in detail by Watson and Ramstad (1987).

Corn Grits

Like head rice, grits are cooked and served in parts of the United States more as a vegetable accompaniment to other menu items than as a main breakfast item, although in meal settings that can include breakfast. Typically made from white corn for table use, grits are usually garnished with butter and salt, not sugar and milk. There is some use in the latter context, but not enough for unequivocal classification as a breakfast cereal. This is consistent with the exclusion of grits from the tabulation in Table 1.

Cornmeal

Meal from white or yellow corn is used in the United States primarily as an ingredient mixed with wheat flour in corn bread, corn muffins, corn sticks, and other comparable delicacies. It can also be cooked as a cereal (cornmeal mush) or as an Italian ethnic menu item (polenta). Some consumers cook it this way first and then allow it to cool and set; it can then be sliced, fried, and eaten like pancakes with syrup.

Corn Flour

As already indicated, corn flour finds use primarily as an ingredient in other food products.

Corn Bran

Although at one time used only in animal feed, with the growing dietary interest in fiber some corn bran is being purified and made available as a fiber-rich food ingredient in other cereals and baked products (Burge and Duensing, 1989). It has not yet found independent application as a hot cereal or hot cereal ingredient.

References

AACC. 1983. Approved Methods of the American Association of Cereal Chemists, 8th ed. The Association, St. Paul, MN.

Babcock, D. 1987. Rice bran as a source of dietary fiber. Cereal Foods World 32:538-539.

Baecker, H. 1987. Processing oats for human consumption. Bull. Assoc. Oper. Millers (April):4939-4942.

Barr, L., Brockington, S. F., Budde, E. F., Bunting, W. R., Carroll, R. W., Gould, M. R., Grogg, B., Hensley, G. W., Rupp, E. G., Stout, P. R., and Western, D. E. 1955. Facts on Oats. The Quaker Oats Co., Chicago, IL.

Billings, H. J. 1938. Method of treating cereals and resulting product. U.S. patent 2,131,881.

Billings, H. J. 1941. Fortified cereal. U.S. patent 2,259,543.

Burge, R. M., and Duensing, W. J. 1989. Processing and dietary applications of corn bran. Cereal Foods World 34:535-538.

Caldwell, E. F., and Pomeranz, Y. 1973. Industrial uses of cereals: Oats. Pages 393-411 in: Industrial Uses of Cereals. Y. Pomeranz, ed. Am. Assoc. Cereal Chem., St. Paul, MN.

Calkins, E. E. 1974. The transition—Magazines into the marketplace in 50 years. Reprinted from Scribner's, January 1937. Advertising Age (Nov. 18):152.

Deane, D., and Commers, E. 1986. Oat cleaning and processing. Pages 371-412 in: Oats: Chemistry and Technology. F. H. Webster, ed. Am. Assoc. Cereal Chem., St. Paul, MN.

Dorn, V. 1989. New machinery in oat milling. Bull. Assoc. Oper. Millers (July):5493-5501.

Gordon, D. T. 1989. Functional properties vs. physiological action of total dietary fiber. Cereal Foods World 34:517-525.

Hoffman, L. A., and Livezey, J., Jr. 1987. The U.S. oats industry. U.S. Dep. Agric. Agric. Econ. Rep. 573.

Huffman, G. W., and Moore, J. W. 1961. Instant oatmeal. U.S. patent 2,999,018.

Johnson, L. 1986. Green oat hulling. Bull. Assoc. Oper. Millers (October):4843.

Juliano, B. O., ed. 1985. Rice: Chemistry and Technology, 2nd ed. Am. Assoc. Cereal Chem., St. Paul, MN.

Karwowski, J. 1985. Method of preparing instant, flaked wheat farina. U.S. patent 4,551,347.

Karwowski, J. 1986. Method of preparing instant, flaked wheat farina. U.S. patent 4,590,088.

Karwowski, J. 1987. Method of preparing instant, flaked wheat farina. U.S. patent 4,664,931.

Marquette, A. F. 1967. Brands, Trademarks, and Goodwill: The Story of The Quaker Oats Company. McGraw-Hill, New York.

Pomeranz, Y., ed. 1988. Wheat: Chemistry and Technology, 3rd ed., Vols. I and II. Am. Assoc. Cereal Chem., St. Paul, MN.

Pomeranz, Y., ed. 1989. Wheat Is Unique. Am. Assoc. Cereal Chem., St. Paul, MN.

Pomeranz, Y., Davis, G. D., Stoops, J. L., and Lai, F. S. 1979. Test weight and groat-to-hull ratio in oats. Cereal Foods World 24:600-602.

Reynolds, J. 1970. Windmills and Watermills. Praeger Publishers, New York.

SAMI-Burke. 1989. Report on sales of all hot cereals (excluding corn grits) in the United States. Arbitron-SAMI, New York.

Seibert, S. E. 1987. Oat bran as a source of soluble dietary fiber. Cereal Foods World 32:552-553.

Sissons, P. 1988. Statement before the Subcommittee on Agricultural Production and Stabilization of Prices, U.S. Senate Agriculture Committee, June 23.

U.S. Code of Federal Regulations. 1988. Title 7, Chapter 810, Part 1004; Title 21, Chapter 137, Parts 300 and 305.

Watson, S. A., and Ramstad, P. E., eds. 1987. Corn: Chemistry and Technology. Am. Assoc. Cereal Chem., St. Paul, MN.

Webster, F. H. 1986. Oat utilization: Past, present, and future. Pages 413-426 in: Oats: Chemistry and Technology. F. H. Webster, ed. Am. Assoc. Cereal Chem., St. Paul, MN.

Wolffing, R. M., Batten, C. J., and Harris, M. C., Jr. 1979. Farina milling process. U.S. patent 4,133,899.

Wood, P. J., Weisz, J., Fedec, P., and Burrows, V. D. 1989a. Large-scale preparation and properties of oat fractions enriched in $(1\rightarrow3)(1\rightarrow4)$-$\beta$-D-glucan. Cereal Chem. 66:97-103.

Wood, P. J., Anderson, J. W., Braaten, J. T., Cave, N. A., Scott, F. W., and Vachon, C. 1989b. Physiological effects of beta-D-glucan rich fractions from oats. Cereal Foods World 34:878-882.

Chapter 10

Fortification and Preservation of Cereals[1]

BENJAMIN BORENSTEIN   LEONARD JOHNSON
ELWOOD F. CALDWELL   THEODORE P. LABUZA
HOWARD T. GORDON

Fortification

Fortification technology must, above all else, be kept pragmatic. After the decisions have been made regarding which nutrients are wanted in the final product and the levels at which they are to be added, the food technologist is faced with the threefold challenge of accomplishing the fortification such that 1) the product is not negatively affected in odor, flavor, or color; 2) the added nutrients are acceptably stable, and sufficient overage is added to compensate for losses in processing and storage; and 3) the process remains practical and economically viable.

The more complex and abusive the processing conditions, the more difficult it is to achieve these ends. However, the demonstrated ability of the cereal industry to routinely fortify millions of tons of breakfast cereal with a wide spectrum of nutrients not only places it in the forefront of fortification technology application but also demonstrates the feasibility of such fortification with available technology and equipment.

In this section, we discuss both the technology decisions that must be made in breakfast cereal fortification and practical means of implementing such fortification.

[1]Principal contributors of the section on fortification, Leonard Johnson, Howard T. Gordon, and Benjamin Borenstein; on preservation, Elwood F. Caldwell and Theodore P. Labuza.

273

RATIONALE

The rationale for the fortification of cereals has been well established. In the United States, the Nationwide Food Consumption Survey (1984) showed that many Americans are not consuming sufficient levels of several nutrients. Zabik (1987), in a further analysis of data from this survey, showed that children, teenagers, and women who consume ready-to-eat cereals at breakfast obtain significantly higher quantities of vitamins and minerals than those not consuming breakfast. Similar studies in other countries would be expected to show even greater nutritional significance of breakfast cereal fortification in improving the nutritional value of the overall diet.

APPROACH TO FORTIFICATION

The general background leading to the determination of which nutrients to add and the levels at which to add them is dealt with in Chapter 11, on cereal nutrition. The restoration of nutrients to cereal foods, usually referred to as enrichment, has been practiced since the 1940s. In the United States, federal standards of identity for the enrichment of wheat flour, bread, and related bakery products were enacted in 1952, and similar standards are in effect for the enrichment of other grain products, including farina, corn grits, cornmeal, and milled rice, as listed in Table 1 (U.S. Code of Federal Regulations, 1988). Enrichment standards for other breakfast cereals have not been set by law, but some products not otherwise fortified to higher levels with a greater variety of micronutrients are restored or enriched at the levels indicated in Table 1.

When it comes to fortification, a prudent approach is recommended. The maximum target for the average consumer's intake of nutrients at each meal should likely be one third of the day's requirement (on the assumption of three meals a day), although it may be practical to fortify breakfast cereals to make available as much as 100% of the daily requirement for most micronutrients in a single serving. Breakfast cereal manufacturers are able to (and do) offer a variety of choices.

Several methods of enrichment or fortification are available. The technique selected must be based on sound manufacturing practices as well as a good knowledge of nutrient chemistry and of the product and the process. Table 2 provides a brief summary of vitamin stabilities that are important for fortification. Cereal product characteristics that must be considered include the desired nutrient claim, product formulation, pH, moisture content, processing temperatures, holding times (if any), storage temperatures, storage times, and packaging.

Knowledge of the product and the nutrient to be added can lead to success of the desired product.

FORTIFICATION TECHNIQUES

The most common technique for fortifying ready-to-eat cereals is to add the minerals and the more stable vitamins (such as niacin and riboflavin) to the basic formula mix and then spray the more labile vitamins (such as vitamin A and thiamin) on the product after

TABLE 1
Levels of Enrichment/Restoration in Some Cereal Grain Products

Component	Enriched Corn Grits[a]	Enriched Corn Meal[a]	Enriched Farina[a]	Enriched Rice[a]	Puffed Wheat and Rice Enriched[b]	Puffed Wheat and Rice Fortified[c]
Thiamin (mg/lb)						
Min.	2.0	2.0	2.0	2.0	1.0 (<2%)	12.0 (25%)
Max.	3.0	3.0	2.5	4.0		
Riboflavin (mg/lb)						
Min.	1.2	1.2	1.2	1.2	··· (<2%)	8.2 (15%)
Max.	1.8	1.8	1.5	2.4		
Niacin (mg/lb)						
Min.	16	16	16	16	12.8 (<2%)	160.0 (25%)
Max.	24	24	20	32		
Iron (mg/lb)						
Min.	13	13	13	13	11.5 (<2%)	144.0 (25%)
Max.	26	26	···	26		
Calcium[d] (mg/lb)						
Min.	500	500	500	500	··· (<2%)	1,280 (4%)
Max.	750	750	···	1,000		
Vitamin D[d] (USP units/lb)						
Min.	250	250	250	250	··· (<2%)	1,280 (10%)
Max.	1,000	1,000	···	1,000		

[a]Amounts in milligrams per pound are U.S. federal standards, from 21 CFR 137.235 for corn grits, 137.260 for corn meal, 137.305 for farina, and 137.350 for rice (U.S. Code of Federal Regulations, 1988).
[b]First entries in this column are calculated amounts of nutrients in milligrams per pound where the ingredient list indicates enrichment, assuming 2% of the U.S. recommended daily allowance (U.S. RDA) per 0.5 oz. serving and based on U.S. RDAs of 1.5 mg for thiamin, 20 mg for niacin, and 18 mg for iron. Entries in parentheses are actual label claims in percent U.S. RDA per 0.5-oz serving for puffed wheat and puffed rice manufactured by The Quaker Oats Co.
[c]First entries in this column are calculated amounts of nutrients per pound where the ingredient list indicates fortification, assuming the claimed percent of U.S. RDA per 0.5-oz serving and based on U.S. RDAs of 1.5 mg for thiamin, 1.7 mg for riboflavin, 20 mg for niacin, 18 mg for iron, 1,000 mg for calcium, and 400 USP units for vitamin D. Entries in parentheses are actual label claims in percent U.S. RDA per 0.5-oz serving for puffed wheat and puffed rice manufactured by Malt-O-Meal Co.
[d]Optional.

processing, by one or another of the Phase I coating methods discussed in Chapter 7. Anderson et al (1976) recommended that vitamins A, D, and C be added to ready-to-eat cereals in a spray solution-suspension after other processing. In their study, the stability of vitamins A and C, both of which are oxygen-sensitive, was enhanced when both were present in the product. Niacin, riboflavin, vitamin B-6, and vitamin E (as the acetate) can be added to the mix before processing, with satisfactory stability. For both the addition of nutrients to the mix before processing and their addition in a spray after processing, a premix or blend of nutrients is generally used, either prepared in-house or purchased from a supplier.

Although technically acceptable, the use of two premixes has its disadvantages. It requires more routine analytic work, since at least one of the mix-added vitamins and one of the sprayed vitamins must continually be checked to confirm both the fact and the level of addition. Furthermore, both premixes must be purchased (or prepared) and inventoried.

A second approach is to spray a suspension of all of the vitamins (obtained as a premix) on the cereal. This technique takes full advantage of the benefits of simplicity in processing (i.e., weighing only one ingredient) and in quality control monitoring, as well as the economy of ordering and stocking only one ingredient rather than two or more.

Premixes produced in-house require even more routine control procedures and create potential problems, since all of the vitamins being

TABLE 2
Sensitivities of Vitamins

	Destabilizing Agents[a]				Optimal pH
	Heat	**Light**	**Oxygen**	**Metals**	
Oil-soluble vitamins					
A	+++	++++	++++	++++	...
D	+++	++++	++++	++++	...
E	++	+++	+++	+++	...
K	+	++++	+++		...
Water-soluble vitamins					
Ascorbic acid	+++	++	++++[b]	+++	2–5
Thiamin	++++	+++	+	+++	2
Riboflavin	+	++++[b]	+	+++	5
Pyridoxine	+	+++	+	+	2
Cyanocobalamin	+	+++	+	++	4–5
Niacin	+	+	+	+	3–7
Pantothenic acid	+++	+	++	+	4–5
Biotin	+	+	+	+	6–7
Folic acid	+	+++	+++	+++	7

[a] + = Stable; ++++ = very unstable.
[b] Unstable in solution but stable in dry form.

added must be bought and inventoried. For example, in order to claim that 100% of the daily allowance of all vitamins is provided by one serving, as many as 12 separate vitamins must be blended into one or two premixes, which must be assayed for potency and uniformity before they are added to the cereal. This requires both in-process and postprocess testing.

A simple variation to markedly simplify the situation is to add just the minerals to the mix and to spray all of the added vitamins on the processed cereal. This allows the purchase of a single blend of vitamins, which may be delivered with a certificate of analysis as part of the quality vendor program, thus significantly minimizing the expense of in-house assays of the premixes.

SPRAY SYSTEMS

Labile vitamins, such as thiamin and ascorbic acid (vitamin C), are typically added by the spray method, to avoid the deleterious effects of pH and temperature on the former and of oxygen and temperature on the latter. Vitamin A is added in the spray to prevent its destruction in processing and to extend its shelf life stability during product storage. For this reason, the composition of the spray is quite important.

The critical factors in designing a spray system are 1) obtaining uniformity in spray coverage and 2) optimizing the protection of vitamins A and D, which pose unique stability problems. Spray coverage is essentially an engineering problem, which appears to have been well resolved by blending in a rotating drum as described in Chapter 7. The hot cereal from the oven enters a rotating drum with stationary spray nozzles at or near its entrance. The spray rate is set in proportion to the volume of hot cereal entering and leaving the drum, to achieve uniform coverage and accurate fortification levels. Chapter 7 provides details on the hardware and control equipment.

PROTECTION AGAINST ATMOSPHERIC OXIDATION

Vitamins A and D cannot be added directly to the cereal mix because of temperature sensitivity. Spraying, however, maximizes their exposure to oxidative degradation. This can be minimized if adequate protection is built into the spray system. Such protection is most readily provided by an appropriate antioxidant system and a carbohydrate oxygen barrier in the spray solution.

A very effective antioxidant system is butylated hydroxytoluene (BHT) or a blend of this with butylated hydroxyanisole (BHA), included in the vitamin mix by the supplier. The simplest and most common oxygen barrier is sucrose, which should constitute at least 10% of the spray formula and is generally used in the 15–25% range. The effects of dextrin and sucrose added to vitamin A spray solutions on vitamin stability in fortified tea leaves were reported by Brooke and Cort (1972).

The data in Table 3 provide evidence that vitamin A stability in ready-to-eat breakfast cereals is directly related to the method of addition and the "coating" protection of added sugar. The data are from an unpublished study by the Hoffmann-LaRoche laboratories of commercially produced cereals in storage for 12 months. The premix contained vitamin A palmitate and sodium ascorbate and also thiamin mononitrate (among other nutrients), included to monitor uniformity. Vitamin A stability was excellent in raisin bran sprayed with a solution containing 15–20% sugar, contrasted with a loss of almost 30% in cornflakes sprayed with a solution containing less than 1% sugar and only BHT to protect the vitamin A. In the case of vitamin C, moisture was probably the key variable in determining stability. A 60% loss occurred in the high-moisture raisin bran, which had been packed at about 6% cereal moisture to minimize raisin hardening during storage, but a loss of only half that occurred in the cornflakes, which had been packed at 2–3% moisture.

What about the ingredient impact of the added water and sugar when all of the vitamins are sprayed on the cereal? In a worst-case scenario, we might assume the goal to be the addition of 100% of the recommended daily allowance (RDA) per 1-oz serving of cereal of all 12 vitamins in the spray. If this were done in the United States with an available stock U.S. RDA vitamin blend and 20% sucrose in the spray solution,

TABLE 3
Vitamin Stability in Cereals as Affected by Sugar in the Vitamin Spray Solution[a]

Cereal Type[b]	Solution Sugar Content[c]	Storage Time (mo)	Vitamin A[d]		Vitamin C[e]		Thiamin[f]	
			IU/oz	Percent Loss	mg/oz	Percent Loss	mg/oz	Percent Loss
Raisin Bran	15–20	0	2,180	...	27.6	...	0.81	...
		6	2,230	0	18.9	22	0.73	10
Cornflakes	< 1	0	1,430	...	18.8	...	0.49	...
		6	1,170	18	17.1	9	0.50	0
		12	1,040	27	13.4	29	0.46	6

[a] Data from an unpublished study of commercially produced cereal stored for 12 months at ambient room temperature, conducted at Hoffmann-LaRoche laboratories.
[b] Cereal type is not intended to denote any particular brand.
[c] Both spray solutions contained butylated hydroxytoluene as an antioxidant.
[d] Vitamin A palmitate.
[e] Sodium ascorbate.
[f] Thiamin mononitrate.

650 mg of spray solution would contain 260 mg of vitamin blend, 260 mg of water, and 130 mg of sugar (sucrose). The spray rate would be only 2.3% (w/w) and would add only 0.46% sugar to the product. The moisture effect would be negligible, since most of the 0.92% water in the spray would immediately evaporate upon being sprayed on the hot cereal just leaving the drying or toasting oven.

OVERAGES

The vitamin or mineral input over and above the label claim is usually referred to as the overage. The amount required is dependent on the intrinsic stability of the vitamin, the food system, the process, the shelf life requirement, the packaging, and analytic error in the concentration of the vitamin in the food.

If the fortification rationale is appropriately conservative, the use of any relatively high overages necessary to ensure meeting the label claim poses no safety issue, with the possible exception of vitamin D. Vitamin D is the most toxic vitamin and probably the most difficult to analyze at food concentration levels. This means that the target fortification level and the overage to be applied must be carefully considered.

The less stable vitamins—thiamin and vitamins A and C—frequently require overages of 50% or higher. Because of physical distribution problems and analytic error, even the most stable vitamins require overages of 15% to minimize problems with the label claim.

Preservation

In view of the heat processing received by most breakfast cereals during manufacture or consumer preparation and their low moisture content during commercial distribution, spoilage as a result of microbial growth is unknown, although not impossible. However, cereals do need to be preserved against shelf life deterioration caused by contact with two other constituents of the atmosphere—water vapor and oxygen. Water vapor uptake may increase the moisture content of ready-to-eat cereals from the 2–3% at which they are usually packed to a level at which they lose their crispness and even (if presweetened by sugarcoating) clump or become sticky. Excess Maillard or nonenzymatic browning can produce undesirable flavors and brown pigments. Hydrolysis of any lipids present may occur if lipolytic enzymes from the original cereal grain have not been inactivated. And any atmospheric oxidation of the lipids in cereals or their ingredients, even to a low extent in a low lipid content, can result in staleness or a rancid aroma or flavor to the point of complete sensory unacceptability, quite apart from any consideration of toxicity.

As will be explained later in discussing atmospheric oxidation, deterioration from moisture and deterioration from oxygen are not independent of each other, even though they may be dealt with separately in product development. As just pointed out and as noted in earlier chapters, achieving the desired crispness in a ready-to-eat cereal usually requires that it be packaged at a moisture content of 3% or less and that little or no moisture gain occur in distribution (shredded wheat being an exception). Maintaining such a low moisture in most climates requires packaging in a liner with a low water vapor transmission rate (WVTR), typically heat-sealed waxed glassine, laminated waxed glassine, or coextruded high-density polyethylene and ethyl vinyl acetate (Monahan, 1988). Details on the packaging of ready-to-eat cereals are given in Chapter 8.

Such liners are not necessary for packaging hot cereals or for some ready-to-eat cereals, including the shredded wheat products made by several companies. Shredded wheat may well be more acceptable at 3% than at 8% moisture, but it remains acceptable at the higher moisture level, and it was in the past packaged in paperboard cartons with paperboard separators and no liners. Now it is typically packed in glassine liners that not only are not laminated but also are generally only lightly waxed, for sealing rather than for barrier purposes. Consumers believed that a stale aroma identified as "paperboard" or "cardboardy" was picked up from the carton or paperboard separators, although this is easily demonstrated to come from incipient atmospheric oxidation of the cereal lipids and not from the package. The nonbarrier single-waxed glassine liner heads off the idea of odor transfer from the carton and at the same time provides a carrier for antioxidants.

MOISTURE GAIN VERSUS SHELF LIFE

Early Studies

Felt et al (1945) determined the shelf life of puffed corn and oat cereals (not otherwise identified, but likely Kix and Cheerios, made by General Mills, Inc.). The cereals were packaged in cartons with waxed glassine liners, stored under atmospheric conditions typical of commercial distribution in 12 U.S. cities, and sampled for moisture content and eating quality every two weeks. The investigators used the following expression to predict the amount of water vapor transmitted through a barrier:

$$W = RAt(p_1 - p_0) , \qquad (1)$$

where W = the weight of the water transmitted, R = the permeability of the barrier, A = the permeable area, t = time, p_1 = the vapor pressure (vp) of the atmosphere with the higher humidity (the storage environment), and p_0 = the vp of the atmosphere with the lower humidity (inside the package). They set out to determine $R_1 = RA$ by holding a packaged product of liner area A under known exact conditions of temperature and humidity and tracking any gain in moisture over time. However, this determination is complicated by the fact that p_0 increases (and thus $p_1 - p_0$ decreases) as the cereal gains moisture. It was thus necessary to use an awkward successive approximation method to obtain a value to use. Even then, it applied to only the particular product and package actually tested, thus imparting a strong element of empiricism to the process.

Having obtained a value of R_1 for a given product and package type, they predicted shelf life from equation 1. It did approximate reasonably well the actual shelf life under the field-test conditions prevailing in most of the 12 cities. The conclusion, still more or less valid for nonsugarcoated cereals , was that a single waxed glassine liner provides an adequate moisture barrier year-round in low-humidity areas (such as Salt Lake City) and for most of the year in areas with low- and high-humidity seasons (such as Chicago), but not at all in high-humidity areas (such as Houston).

In a more rigorous and basic study, Heiss (1958) investigated the relationship between the WVTR of the packaging film, the moisture sorption properties of foods, and the resulting shelf life. He applied engineering concepts of mass transfer and dealt with such additional factors as pinholes in the film, moisture distribution in the food, and the effects of the environment on film permeability. However, at that time, solutions to the equations involved still had to be done by hand calculation using numerical integration, which limited their practical usefulness in product and package development. Karel and others later applied the same principles but solved the model originally proposed by Heiss using numerical integration on a computer and assuming the isotherm of moisture versus water activity to be linear over the region involved (Karel, 1967; Mizrahi et al, 1970; Labuza et al, 1972).

Moisture Sorption Isotherms

All foods have characteristic curves of moisture versus equilibrium relative humidity or water activity (a_w) at a given temperature; these curves are known as moisture sorption isotherms, illustrated in general form in Figure 1. The portion applicable to crisp ready-to-eat cereals is generally below $a_w = 0.50$. Starting with Salwin (1959), various investigators have proposed equations to approximate such isotherms based on the Brunauer-Emmett-Teller (BET) sorption theory that a monomolecular layer of water molecules is initially formed on the polar sites of a food surface before additional layers are begun. Karel and Labuza (1969) proposed a linear equation found to be applicable in the a_w range from the monolayer value up to 0.6. The key observation from the BET equations is that the maximum shelf life of most dry foods (including cereals) in terms of nonenzymatic browning and atmospheric oxidation is at the moisture content or water activity corresponding to the monolayer value. It is also the case that if a cereal is sugarcoated, moisture gain to an a_w value above 0.4–0.5 results in stickiness and clumping (Labuza, 1984b). According to Taoukis et al (1988), a more complete equation based on the BET theory and applicable to most foods over a wider range of a_w is the Guggenheim-Anderson-DeBoehr (GAB) isotherm equation.

The linear equation is

$$m = ba_w + c , \qquad (2)$$

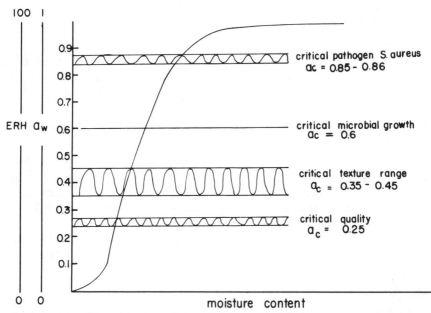

Figure 1. General sorption isotherm of water activity (a_w) versus moisture content, showing critical water activities (a_c). ERH = equilibrium relative humidity. *S. aureus = Staphylococcus aureus*. (Reprinted, with permission, from Labuza and Contreras-Medellin, 1981)

where m = the predicted moisture content, b is a constant reflecting the slope of the isotherm for a_w between 0.2 and 0.6, and c is the intercept, or the moisture level extrapolated to $a_w = 0$. The GAB equation is

$$m = \frac{m_0 C k a_w}{[1 - k a_w)(1 - k a_w + C k a_w)]}, \tag{3}$$

where m_0 = the monolayer moisture value, C is a constant relating to interactive energy between the water and the food, and k is a constant relating to interactive energy between multiple layers of water. For $k = 1$, the GAB equation reduces to the BET equation

$$m = \frac{m_0 C a_w}{(1 - a_w)(1 - a_w + C a_w)}. \tag{4}$$

A plot of $a_w/[(1 - a_w)m]$ for various values of a_w up to 0.5 gives a straight line. From this, the monolayer moisture is $m_0 = 1/(\text{intercept} + \text{slope})$.

From these isotherms, it is also possible to determine the direction of moisture flow among ingredients during storage. Those with a higher a_w or vp lose moisture to those of lower vp. For example, a raisin at $a_w = 0.6$ gives up moisture to cereal at $a_w = 0.1$ in the same package, and they eventually reach the same a_w, at which point the raisin may be inedibly hard. Similarly, if the outside humidity is high, the atmospheric vp is high, driving water vapor into the package and causing the cereal a_w to go above the optimum value for maximum shelf life.

Package Barrier Considerations

Labuza and others reviewed the determination of the moisture protection requirements of foods (Labuza and Contreras-Medellin, 1981; Taoukis et al, 1988), confirming that the control of moisture exchange with the environment by packaging is crucial for moisture-sensitive foods, and presenting models for the prediction of moisture transport based on moisture sorption isotherms. Taoukis et al (1988) presented prediction models based on the linear isotherm (equation 2) and the GAB isotherm (equation 3). This analysis is based on the more rigorous treatment of the moisture transfer equation

$$\frac{dW}{dt} = \frac{k}{x} A (p_1 - p_2), \tag{5}$$

where W = the weight of water transferred, k/x = the permeability of the film, A = the package area, p_1 = the outside vp, and p_2 = the inside vp in equilibrium with the cereal, assuming constant external relative humidity (RH) and temperature. Since a_w = (% RH)/100 = p_2/p_0, where p_0 = the vp of pure water at temperature T, the end result by the linear isotherm is

$$\ln \Gamma = \ln[(m - m_i)/(m_e - m_i)] = (k/x)(A/W_s)(p_0/b)t , \qquad (6)$$

where Γ = the unaccomplished moisture change, m = the moisture content (%, db) at time t, m_e = the final equilibrium moisture content (%, db) reached by the cereal on the linear isotherm at the assumed external RH, m_i = the initial moisture content (%, db), W_s = the dry weight (g) of the cereal in the package, and b = the isotherm slope ([g water]/[g dry solids]).

Thus, a plot of $\ln \Gamma$ versus time (and not moisture content or water gain versus time) is a straight line, as was originally shown by Mizrahi et al (1970). An analytic solution that is much more complete exists for the GAB isotherm. Several computer programs exist for solving these equations (T. P. Labuza, personal communication).

An example of these calculations for cornflakes appears in Figure 2, showing that the moisture contents predicted by the models are in quite good agreement with those measured experimentally. Thus the shelf life of a product within a given barrier material can be predicted from the maximum tolerable product moisture and the permeance of the barrier (both of which can be determined experimentally) and the RH of the storage environment (which can be estimated or assumed).

One key aspect of this solution is that it can be quickly used to see the effect of different temperatures, RHs, and ratios of area to weight (or area to volume). For example, it can be shown that the time taken to reach a given critical moisture content in a cereal in a small, individual serving carton is about half that in the typical family-size carton (which has a smaller A/W_s). Thus, an overwrap is needed to lower k/x for individual cartons such as those used with six- or eight-pack trays (Taoukis et al, 1988). In addition, if a time-temperature-humidity distribution is known, one can accurately calculate the moisture versus time under these variable conditions by breaking the solution up into short time sequences and solving for each of these times. This was shown to work well for packaged wild rice (Gencturk et al, 1986) and pasta (Cardozo and Labuza, 1983).

Permeance

The WVTR (in g·day^{-1}·m^{-2}) and the permeance to moisture vapor (k/x, in g·day^{-1}·m^{-2}·mmHg^{-1}) of an intact package film can be determined by method E-96 of the American Society for Testing and Materials (ASTM, 1980) by sealing a desiccant-containing cup with the film and recording weight gain while holding the cup at 100°F (37.8°C) and 90% RH. A similar practical measurement can be made using the package itself containing the desiccant. The desiccant maintains 0 vp internally. The vp gradient across the film is thus constant at 44.3 mmHg, since the vp of pure water (100% RH) is 49.3 mmHg at 100°F. Over a 7- to 10-day period a constant weight gain should be achieved. Thus dw/dt = (g water)/day = WVTR × A, where A = the area of the film, or WVTR = S/A, where S = the slope of water gain versus time.

To account for different vp gradients and solve for the condition where the vp inside changes as moisture is gained, the permeance is calculated as k/x = WVTR/ΔP (in [g water]·day^{-1}μm^{-2}·mmHg^{-1}), with ΔP being the vp gradient across the film.

WVTR as used in the packaging industry and as given in the cereal carton liner specification in Chapter 8 may use the same determinative

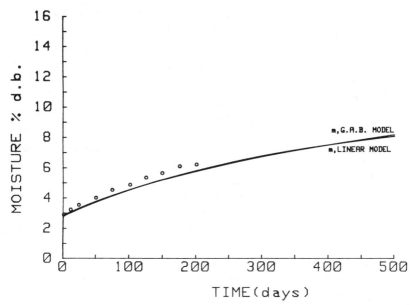

Figure 2. Moisture gain of cornflakes packaged in polystyrene. The points are experimental data, and the continuous curves represent predictions from packaging models, in this case almost identical to about 8% mosture. (Reprinted, from Taoukis et al, 1988, with permission from the American Chemical Society)

methodology but omits the correction for the vp gradient. Its use in packaging studies thus may involve incorrect assumptions. Permeance, on the other hand, is a property of the film. It depends on the nature of the film, its thickness, the temperature at which it is measured, and the external RH. Values have been compiled in tables (Bakker, 1986) and may also be provided by package material manufacturers, but actual permeance in a package may be affected by the packaging process, including the effect of poor seals, pinholes, and creases. Thus redetermination using the formed package (containing a desiccant and held at the specified temperature and humidity) instead of the official cup may be advisable.

The permeances (in [kg water]$\cdot$day$^{-1}\cdot$m$^{-2}\cdot$mmHg^{-1}) of a range of packaging materials measured at 95°F (35°C) and 100% RH are as follows (Taoukis et al, 1988): paperboard, 3,333; polypropylene, 137; cellophane/polyethylene, 102; polyethylene (1 mil), 86; polyethylene/terephthalate, 50; polyester, 33; polyethylene/terephthalate/polyethylene, 19; and polyester/foil/polyethylene, 1.

In practice, once advance projections of shelf life have been made on the basis suggested here and a package liner has been chosen and successfully field-tested (say, during test marketing), it should not be necessary to repeat all tests for similar products going into the same market areas with similar packaging.

NONENZYMATIC BROWNING

Another critical parameter affecting cereal quality is the extent of the Maillard browning reaction occurring during processing and distribution. This was reviewed by Saltmarch and Labuza (1982), Labuza and Schmidl (1986), and Kaanane and Labuza (1988). Specifically, the reaction occurs between reducing compounds (especially sugars added in cereal formulations, such as dextrose, fructose, and lactose) and amines (such as the lysine naturally present in cereal proteins or added in the form of dairy or soy proteins). The main products of the reaction include both desirable and undesirable cereal flavors and brown pigments. Like other reactions, Maillard browning proceeds at a rate that is directly influenced by water activity, going to zero at a_w below the monolayer level and reaching a maximum at a_w in the range of 0.6–0.8.

One way to control browning is to minimize the amount of reducing sugars. This is generally not economical. The second is to ensure good control of the a_w in processing and through packaging. Inhibitors such as sulfite and cysteine may be effective but can contribute to off-flavors and also are of questionable regulatory status. As pointed out later in discussing lipid oxidation, some browning products actually have an antioxidant effect. In general, browning follows a linear relationship with time, but because of the many interleaved reaction paths, much less is understood of the mechanisms of the production of desirable or undesirable flavors.

HYDROLYTIC RANCIDITY

It was mentioned earlier in this chapter that the hydrolysis of lipids in a food can occur as a result of lipolytic enzyme action. The release of free fatty acids can result in an undesirable flavor and aroma. This effect has been termed hydrolytic rancidity, perhaps by analogy with rancid butter, the aroma of which results directly from the release of some of the volatile short-chain fatty acids in milk fat, particularly butyric acid. Most cereal lipids have longer-chain, less volatile fatty acids, whose presence in free form in a food results in a bitter or soapy flavor rather than a rancid odor.

Cereal lipases are nonspecific and can attack not only the cereal lipids but also other fats in the formulation. This applies not only to such formulated products as baking mixes but also to breakfast cereals containing added fats or oils for flavor, texture, or other functional purposes. Examples include a ready-to-eat cereal sprayed with a sugarcoating containing fat and an instant hot cereal to which a coffee whitener type of mix has been added, containing an emulsified vegetable oil such as coconut oil, the lauric acid content of which would be expected to give a soapy flavor if hydrolyzed to the free fatty acid or acid salt form.

Lipolytic enzymes in a cereal grain such as oats are released when the structure of the kernel is disrupted by cutting, rolling, or other grinding. the greater the moisture content of the grain when this occurs, the faster the enzyme action, and the greater its effect. Grogg and Caldwell (1958) noted that the action can go far enough in just a few days to make the oat flour from green (i.e., non-heat-treated) oat groats behave rheologically as though it had been defatted.

Water activity is a major factor controlling residual lipase activity, which is generally at a minimum in storage if the moisture level is below that of the monolayer, and it increases as the product gains moisture (Acker, 1969).

It would be difficult to routinize a direct test for lipase. However, color tests based on other marker enzymes, such as tyrosinase or peroxidase, are available. Procedures for qualitative tyrosinase and peroxidase tests are given in the accompanying box. Quantitative versions, as yet unpublished, are also in use. The sparging steam normally applied to whole or cut oat groats before rolling should be a required operation and should produce a negative result by either of these tests. However, rolled oats and fractions such as oat bran and whole or refined oat flour might receive insufficient steaming or bypass the steaming operation, and products of unknown history should be tested before use.

OXIDATIVE RANCIDITY

Breakfast cereals were traditionally as low in fat as the grains from which they were made, or even lower. Fat contents lower than that of the grain itself could result from milling operations, such as the removal of the germ from wheat and corn before manufacturing cereal products and the pearling of rice kernels before gun puffing. However, low fat content is no guarantee against flavor deterioration due to atmospheric oxidation of even the minimal fat content of processed cereal products, since very little of the fat needs to oxidize to result in organoleptic rancidity (Labuza, 1971). Also, many breakfast cereals are higher in fat than their grain ingredients, if fat-containing flavoring and texturizing ingredients have been added.

Mechanism of Lipid Oxidation

Most natural plant lipids contain in their hydrocarbon chain unsaturated linkages susceptible to atmospheric oxidation, a chain reaction that can be triggered by such catalysts as light, heat, metallic ions, and stray free radicals, as well as water activity either higher or lower than the monolayer value. The lipid oxidation reaction involves several simultaneous stages—an *induction* stage, in which free radicals are continuously being formed from the lipid; a *propagation* stage, in which the lipid free radicals oxidize to form hydroperoxides; a *termination* stage, in which some of the free radicals are trapped by antioxidants, proteins, or other radicals; and a *disintegration* stage, in which the hydroperoxides break down to form off-flavors.

In the course of the reaction, oxygen consumption (as well as lipid disappearance) is slow at first, sometimes incorrectly called the induction period. Peroxide levels measured during this time are low and variable. In the next stage, oxygen consumption increases rapidly, and peroxides begin to accumulate. At some point peroxide degradation becomes so rapid that the peroxide value may fall to zero. During all this time, volatile short-chain breakdown products from the peroxide

Qualitative Tests for Enzymes in Oat Products[a]

Tyrosinase

Apparatus: Balance, capable of weighing to 0.1 mg; 20-ml pipette; 1-ml pipette; 50-ml graduated cylinder; centrifuge tube or test tube with stopper; centrifuge.

Reagents: Catechol solution: Weigh 1 g of catechol (pyrocatechin) (Fisher P 370) into 20 ml of distilled water.

Procedure: 1. Weigh 2 g of ground sample; mix with 30 ml of distilled water in a centrifuge tube or test tube; shake all samples about equally to keep oxygen content present and even.
2. Add 1 ml of catechol solution; shake, and let stand for 20 min.
3. Centrifuge for 10 min (may be omitted).

Results: If no significant pink color is seen in 20 min, the result is considered negative. This can be determined roughly without the centrifuging step.

Peroxidase

Reagents: 1. Araboascorbic acid: Dissolve 100 ml of araboascorbic acid in 100 ml of distilled water in a 100-ml volumetric flask. Store in refrigerator. Make fresh solution weekly.
2. Dichloroindophenol: Dissolve 40 mg of dichloroindophenol in 200 ml of distilled water in a 200-ml volumetric flask. Store in refrigerator. Make fresh solution monthly.
3. Hydrogen peroxide: Dissolve 4.0 ml of hydrogen peroxide (30%) in 100 ml of distilled water in a 100-ml volumetric flask. Store in refrigerator. Make fresh solution twice weekly.

Procedure: Place 1 g of ground oat product in a test tube. Add 50 ml of distilled water (room temperature), 2 ml of araboascorbic acid solution (2-ml volumetric pipette), and 3 ml of dichloroindophenol solution (2-ml volumetric pipette). Tightly close test tubes and shake until dissolved. Place test tubes in water bath at 38°C and set timer for 5 min. After 5 min, check for color change. Shake tubes and run again; set timer for 5 min.

Results: After 10 min, if no color change is seen, the result is considered negative. Faint blue indicates low peroxidase; light blue, medium peroxidase; dark blue, high peroxidase.

[a]Tyrosinase qualitative test from National Oats Co., Cedar Rapids, IA; peroxidase qualitative test from ConAgra, Inc., Omaha, NE. In both cases, quantitative procedures are also available (unpublished).

(such as hexanal) accumulate in the package, which can lead to an off-odor if the critical concentration is reached. Generally this occurs after the first stage but before the peroxide value reaches a maximum. A plastic barrier film may scalp (i.e., absorb) the volatile, or a paperboard barrier may let it pass out of the package, so that the cereal may become extensively oxidized before it develops an off-odor. Generally less than 1% of the unsaturated lipid needs to be oxidized for off-odors to be detected.

Catalysts such as light, high storage temperatures (30–50°C), and trace metals (such as iron used in enrichment or adventitious copper or manganese) increase the oxidation rate, by speeding up the first step of free radical formation, and thus shorten shelf life. And, as noted earlier, moisture gain or loss above or below the monolayer value also can speed oxidation, by a complex process.

Relationship to Water Activity

Labuza and co-workers have studied the chemical kinetics of food deterioration (including lipid oxidation) intensively and extensively for many years (Labuza, 1971, 1980, 1982, 1984a; Labuza and Taoukis, 1990). The other catalysts already mentioned (and discussed later) are certainly involved, but the water activity of a lipid-containing food is a key determinant of its likely shelf life in terms of organoleptic (and therefore likely oxidative) rancidity. As mentioned earlier in discussing sorption isotherms, this idea was introduced by Salwin (1959), who showed that for most foods, an a_w of 0.2–0.3 corresponds to a monolayer of water molecules on the food surface interfacing with the surrounding oxygen-containing atmosphere, and that this is the optimum condition for stability against oxidation. The moisture level associated with this a_w range varies with different foods, and at levels above or below it they deteriorate faster.

One significance of this is that it dispelled the early notion (still held by some) that the drier a low-moisture or dehydrated food is, the more stable it is. Unfortunately for some dry ready-to-eat cereals, the most stable point as determined from the sorption isotherm is above the optimum level for crispness. When this is the case, some compromise is involved, and other means of extending shelf life, such as the addition of antioxidants, may have to be invoked, if natural stability is inadequate. Shredded wheat is an example: its isotherm indicates that the monolayer point is well above the 3% moisture range targeted for packaging puffed wheat and many other ready-to-eat cereals. Toasting to 3% moisture may enhance the initial flavor, but shredded wheat is still very acceptable at 7–8% moisture and may reach this range before consumption anyway, with the packaging usually used. There is thus little reason to dry it to 3% moisture, and there are at least two good reasons not to (loss of stability and loss of yield).

Thus an important element in preserving a packaged cereal against deterioration due to atmospheric oxidation is determination of its sorption isotherm and optimization of its moisture content for maximum stability, if possible in view of other desired product characteristics that are also influenced by moisture content. However, the question remains as to how to predict shelf life, once the initially desired formulation, processing, and packaging have been determined and adjusted.

SHELF LIFE AND SHELF LIFE TESTING

Shelf life is foreshortened in different ways, four of which have already been discussed—loss of crispness and flavorability due to water uptake by hygroscopic products; excessive nonenzymatic browning; the so-called hydrolytic rancidity due to enzyme release of free fatty acids; and rancidity due to oxidative decomposition along the hydrocarbon chains of fatty acids in triglyceride form. Each of these has been discussed primarily in terms of sensory acceptance, which is certainly of paramount importance. However, two other forms of deterioration, against which preservation might conceivably be required, are spoilage or toxicity due to microbial action and loss of nutritive value in general (and of label compliance in particular). As pointed out earlier, the first of these is unlikely enough to warrant much attention in a book such as this, although this is not in any way intended to diminish the necessity of following good manufacturing practices and maintaining strict standards of sanitation for incoming ingredients and plant operations. Both of these matters have been dealt with in other books, including three published by the American Association of Cereal Chemists (Christensen, 1982; Baur and Jackson, 1982; Baur, 1984).

Since all breakfast cereals (at least in the United States) display nutritional labeling, which must show all significant nutrients, the preservation of nutritional value and maintaining label compliance in the distribution system are essentially one and the same thing. Accordingly, one or more vitamins known to be of lower stability than the others present or added should be assayed, along with moisture and sensory data, in any shelf life prediction.

Accelerated Testing

In addition to basic study of the kinetics of food deterioration, accelerated shelf life testing (ASLT) has been a related area of study by the Labuza group at Minnesota. ASLT has been practiced in one form or another for at least 65 years. One of the earliest forms of it resulting in any published data was the Schaal oven test, introduced about 1923 for cookie and cracker shortenings (Joyner and McIntyre, 1938). A 50-g sample of shortening or a quantity of a standard cookie or cracker baked with it is placed in a covered beaker or jar in an air oven at 145°F (63°C), and the container is opened briefly and sniffed each day until rancidity is detected. Variations of this have been practiced (and perhaps still are), such as for testing baked products at various temperatures in the 100–140°F (38–60°C) range, including peroxide determinations in an attempt to be more quantitative. However, apart from lack of precision, the flaw in all such tests is that the relationship to actual shelf life is either unknown or, if determined empirically, different for different products. Among other things, reactions may occur or contacts between molecules may be made at the higher temperature that would not occur at all during normal shelf life, such as those resulting from melting and wicking or migration of a fat that would otherwise remain solid. Also, nonenzymatic browning tends to be the predominant mode of failure above 35°C, whereas lipid oxidation predominates below that temperature.

Accelerated tests have generally involved a presumed or empirically determined degree of acceleration by an arbitrarily elevated temperature, with various chemical as well as sensory measurements of lipid oxidation. Caldwell and Grogg (1955) measured the red color developed at 532 μm by thiobarbituric acid reacting with cereal lipid oxidation products. Stuckey (1955) stored cereals (treated in various ways) at 74 and 145°F (23 and 63°C), and conducted periodic organoleptic panel evaluations. Caldwell and Shmigelsky (1958) stored packaged shredded wheat at 100°F (38°C) in unlined cartons and showed that shelf life could be extended by adding BHA to the paperboard of the carton dividers. Anderson et al (1963) stored cereals (treated in various ways) in double-waxed glassine at 37°C (99°F) and in glass jars at 57°C (135°F) and periodically determined peroxide values and evaluated organoleptic quality. Shmigelsky et al (1964) relied upon hedonic rating panels with statistical analysis in comparing package treatment with direct treatment of a shredded oat cereal with BHT in terms of their effects on the shelf life of cereals stored at room temperature and at 100°F (38°C).

Anderson et al (1976) stored various cereals at "room temperature," said to be about 72°F (22°C), at 104°F (40°C), and in a "weather room" in which temperature and humidity were cycled "to simulate day and night conditions in warm coastal cities." Losses of micronutrients were tracked by chemical and instrumental analysis.

None of these tests was predictive in other than perhaps a relative way among the variables in any one series. However, since 1979, the work of Labuza on the kinetics of food deterioration has been extended and applied specifically to ASLT (Ragnarsson and Labuza, 1977; Labuza, 1979; Labuza and Riboh, 1982; Labuza and Kamman, 1983; Labuza, 1985; Labuza and Schmidl, 1985, 1988). Labuza and Schmidl (1985) gave a primarily practical and applied account of ASLT, and a chapter by Labuza (1985) contains a clear explanation of test design using kinetics. Labuza and Schmidl (1988) dealt specifically with using sensory data as the quality criterion; the other studies cited make use of other, more objective attributes, such as moisture gain or chemical changes (e.g., the production of hexanal).

PRODUCT DEVELOPMENT CONSIDERATIONS

The answers to providing adequate shelf life implied in the discussion of moisture gain in breakfast cereals were either 1) to formulate products at a level of hygroscopicity appropriate to an existing or otherwise affordable package barrier or 2) to select, on the basis of permeance testing, a barrier adequate to protect the product through the assumed distribution conditions. In either case, the alternatives of shifting to a different product entirely, or to no product, or to a shorter distribution time, or to marketing in a different climate are assumed to be inherently available and need no discussion here.

We have discussed at some length the kinetics of food deterioration by browning and atmospheric oxidation, with emphasis on the role of water activity and on sounder, faster, and more accurate prediction of the shelf life of a given product. If these steps fall short of our needs, what remains to be done? The basic answers clearly include one or more of the following: 1) scavenge oxygen from the package atmosphere or replace it with an inert gas, 2) exclude light, 3) reduce or eliminate active traces of metals known to catalyze oxidation, 4) formulate and process the product so as to provide natural antioxidants or replace susceptible fats, and 5) add browning inhibitors or chemical antioxidants of various kinds to the food.

Oxygen Elimination

Neither scavenging oxygen with an oxygen absorber such as glucose or catalyzed iron oxidase nor eliminating it by evacuation and replacement with nitrogen seems practical in the usual cereal carton and liner, nor has such a combination been shown to be feasible in any published literature. However, Sakamaki et al (1988) showed that a catalyzed iron oxygen absorber packet can retard or delay lipid oxidation in an oat cereal when combined with a good oxygen barrier material, such as polypropylene polyethylene coated with polyvinylidene chloride. If the cereal were packed in a rigid fiber or metal can and sold, say, for export or even domestically as a specialty item, so as to be able to recover the cost, then one or both of these approaches might be feasible.

Excluding Light

Light, especially the ultraviolet wavelengths, is known to catalyze oxidative reactions. Most cereal packages exclude light for reasons other than stability, such as mechanical protection, esthetic appearance, and the need for space for package copy and illustration. However, anyone planning for any reason to package an oxidation-sensitive product in, say, clear see-through plastic would be well advised to reconsider.

Avoiding Trace Metals

Some heavy metals in trace quantities promote lipid oxidation, particularly copper, manganese, and iron; chromium, nickel, and vanadium are next in importance (Bailey, 1951). Traces of these are found in cereal grains, package materials, and some ingredients. They also enter from process equipment such as mixers, rolls, kettles, piping, valves, and fittings, or as nutritional additives. In known incidents (although not described in published reports), copper present only in the construction of a control valve and manganese in parts per million in the butyl rubber in the wax of a waxed paper liner were identified as rancidity promoters. In other circumstances that would have been expected to provide relatively much greater traces of the same metals in similar products, no comparable rancidity was observed. Thus, it seems as though the interaction is a complex one, perhaps depending upon some kind of activation of the trace metal or of receptor sites or autogeneration of a natural oxidant or antioxidant reaction of an unknown nature.

In any event, it would seem prudent to avoid any use of copper in processing equipment or lines, to use ceramic or plastic or stainless steel wherever possible and feasible, and to minimize metal ions in formulation, especially copper and manganese.

Formulation and Processing

Although Maillard browning in cereals reduces protein quality, it may also generate reaction products that have antioxidant properties.

Results obtained by Anderson et al (1963) appeared to indicate that a certain amount of browning favorably affects the storage stability of flaked cereals. Wheat flakes, oat flakes, and cornflakes that had been cooked and toasted were more stable than controls made by the same formula but freeze-dried or hot-air-dried rather than toasted after cooking. (On the other hand, freeze-dried flakes of raw wheat, oats, and degerminated corn were even more stable than the cooked and toasted flakes.) Lingnert (1980) and Lingnert and Waller (1983) attributed lipid antioxidant properties to the reaction products of glucose and histidine.

Salt has been observed to reduce the storage stability of processed cereal products (Lindeman, 1953; E. F. Caldwell, unpublished). In gun puffing certain grains, particularly oats, an immediate destabilization occurs with even the lightest possible puffing, whereas more drastic preheating and puffing results in a relatively more stable product (E. F. Caldwell, unpublished).

What these observations tell us is that the basis of stability versus susceptibility to atmospheric oxidation is a complex one as related to product composition and processing. Natural prooxidants and antioxidants may both be present and may interact differentially with ingredients and processing. Before any change is made in formula or processing, its possible effect on stability should be considered and tested for. A product with a satisfactory shelf life may or may not be successful for the reasons that we think it is.

Antioxidants

Antioxidants should not be added unless shown to be required. They have been known since the 1950s to be of proven effectiveness, and a number of the publications already cited in discussing the variety and nonuniformity of methods of testing shelf life demonstrate this. Stuckey (1955) showed a cereal shelf life, as measured by the Schaal 145°F (63°C) oven test, to be tripled as a result of spraying on a combination of BHA, BHT, propyl gallate (PG), and citric acid in an oil and water emulsion. Anderson et al (1963) obtained modest increases in the shelf life of wheat flakes, puffed oats, and 40% bran flake cereals and sharper increases in the shelf life of an oat and wheat germ cereal, cornflakes, and raw 40% bran flakes by spraying them with a BHA-BHT mixture, but similar additions of α- and γ-tocopherols had essentially no effect. As pointed out earlier, Caldwell and Shmigelsky (1958) extended the shelf life of shredded wheat by packaging it in unlined paperboard sprayed with BHA in a vegetable oil, and Shmigelsky et al (1964) found that BHT could appreciably extend the shelf life of a shredded oat cereal, whether added as a component of the wax on the glassine package liner or dissolved in vegetable oil and sprayed directly on the product.

In more recent work, Dougherty (1988) and Megremis (1990) showed, contrary to the earlier findings of Anderson et al (1963), that mixed tocopherols may have some effect in extending the shelf life of cereals. Coulter (1988) reviewed the role and effectiveness of the phenolic antioxidants BHA, BHT, and tertiarybutyl hydroquinone (TBHQ), giving detailed information about each, including its structure, chemical properties, and maximum usage levels permitted in the United States.

A term that used to be heard in discussion of the effectiveness or lack of effectiveness of antioxidants is "carry-through." Some antioxidants, notably PG, are very effective in stabilizing shortening, but the stability imparted fails to carry through to baked products in which it is used, because of destruction, volatilization, or steam distillation of the antioxidant during baking. The same failure of carry-through also occurs in cereals. This favors the addition of antioxidants by spraying the product after extruding, baking, or toasting or by allowing transfer over time from the package material.

At the time of writing (1989), mixed tocopherols, BHA, BHT, TBHQ, and PG are all considered to be generally recognized as safe (GRAS) by the U.S. Food and Drug Administration (with a limitation on BHA, BHT, TBHQ, and PG of 200 ppm of the fat or oil content of the food), and all are approved by the Joint Expert Committee on Food Additives of the Food and Agriculture Organization and the World Health Organization. No specific GRAS affirmation by the Food and Drug Administration has been published for tocopherols. In addition, a special food additive regulation in the United States allows BHA or BHT or a combination thereof in concentrations of up to 50 ppm in dry breakfast cereals, calculated as a proportion of the cereal, regardless of its fat content. This is important because otherwise the low fat content of most cereals would place the GRAS amount of antioxidant so low as to make its effective incorporation operationally impossible.

BHA, BHT, TBHQ, or a combination dissolved in a vegetable oil and sprayed on after baking or toasting is a typical direct treatment where an antioxidant is needed. Some synergism may exist between BHA and BHT and possibly between them and citric acid. Antioxidant manufacturers can provide these compounds or premixed combinations already dissolved in a vegetable oil carrier.

Hoojjat et al (1987) recently published results indicating significant transfer of BHT to a flaked oat cereal packaged and stored in high-density polyethylene pouches in which the BHT had been incorporated. Oxidation of the cereal lipids, as followed by a modification of the thiobarbituric acid test of Caldwell and Grogg (1955), appeared to be retarded; no sensory results were reported. Much earlier, Caldwell et al (1964) reported significant BHT transfer to a shredded oat cereal from the wax in a waxed glassine liner, with a favorable effect on poststorage sensory acceptance. For package material application, working with a converter to provide a package liner containing BHT in the wax or plastic is recommended. Significant transfer to the product from the liner may take several days or weeks. Testing the cereal product for compliance with applicable regulations is recommended.

Either mode of addition requires an ingredient listing such as "BHA added to preserve freshness" or "BHT added to package material to preserve freshness." As indicated previously, tocopherols may be preferred because of their natural origin and relationship to vitamin E. However, they are not as effective as the phenolic products on a weight or cost basis. In a recent informal survey, 71 of 122 brands of ready-to-eat cereal in a large U.S. super warehouse market listed BHA, BHT, or TBHQ as an ingredient in the product or package material. Among hot cereals, the proportion was 10 out of 28 brands.

References

Acker, L. 1969. Water activity and enzyme activity. Food Technol. 23:27.

Anderson, R. H., Moran, D. H., Huntley, T. E., and Holahan, J. L. 1963. Responses of cereals to antioxidants. Food Technol. 17:1587.

Anderson, R. H., Maxwell, D. L., Mulley, A. E. and Fritsch, C. W. 1976. Effects of processing and storage on micronutrients in breakfast cereals. Food Technol. 30(5):110.

ASTM. 1980. Methods of the American Society for Testing and Materials. The Society, Philadelphia.

Bailey, A. E. 1951. Industrial Oil and Fat Products, 2nd ed. Interscience, New York.

Bakker, E. 1986. Encyclopedia of Packaging Technology. John Wiley & Sons, New York.

Baur, F. J., ed. 1984. Insect Management for Food Storage and Processing. Am. Assoc. Cereal Chem., St. Paul, MN.

Baur, F. J., and Jackson, W. B., eds. 1982. Bird Control in Food Plants. Am. Assoc. Cereal Chem., St. Paul, MN.

Brooke, C. L., and Cort, W. M. 1972. Vitamin A fortification of tea. Food Technol. 26(6):50.

Caldwell, E. F., and Grogg, B. 1955. Application of the thiobarbituric acid test to cereal and baked products. Food Technol. 4:185.

Caldwell, E. F., and Shmigelsky, S. 1958. Antioxidant treatment of paperboard for increased shelf life of package dry cereal. Food Technol. 12:589.

Caldwell, E. F., Nehring, E. W., Postweiler, J. E., Smith, G. M., Jr., and Wilbur, C. H. 1964. Package treatment versus direct application as a means of incorporating BHT in shredded breakfast cereals. Food Technol. 18:125.

Cardozo, G., and Labuza, T. P. 1983. Effect of temperature and humidity on moisture transport for pasta packaging. Br. J. Food Technol. 18:587.

Christensen, C. M., ed. 1982. Storage of Cereal Grains and Their Products, 3rd ed. Am. Assoc. Cereal Chem., St. Paul, MN.

Coulter, R. B. 1988. Extending shelf life by using traditional phenolic antioxidants. Cereal Foods World 33:207.

Dougherty, M. E., Jr. 1988. Tocopherols as food antioxidants. Cereal Foods World 33:222.

Felt, C. E., Buechele, A. C., Borchardt, L. F., Koehn, R. C., Collatz, F. A., and Hildebrand, F. C. 1945. Determination of shelf life of packaged cereals. Cereal Chem. 22:261.

Gencturk, M. B., Bakshi, A. S., Hong, Y. C., and Labuza, T. P. 1986. Moisture transfer properties of wild rice. J. Food Process Eng. 8:243.

Grogg, B., and Caldwell, E. F. 1958. Gelatinization of starchy materials in the farinograph. Cereal Chem. 35:196.

Heiss, R. 1958. Shelf life determinations. Mod. Packag. 31(8):119.

Hoojjat, P., Hernandez, R. J., Giacin, J. R., and Miltz, J. 1987. Mass transfer of BHT from high density polyethylene film and its influence on product stability. J. Packag. Technol. 1:78.

Joyner, N. T., and McIntyre, J. E. 1938. The oven test as an index of keeping quality. J. Am. Oil Chem. Soc. 15:184.

Kaanane, A., and Labuza, T. P. 1988. The Maillard reaction in foods. Pages 301-328 in: Aging, Diabetes, and Nutrition. J. Baynes, ed. A. R. Liss Press, New York.

Karel, M. 1967. Use tests—Only real way to determine effect of package on food quality. Food Can. 27:43.

Karel, M., and Labuza, T. P. 1969. Optimization of protective packaging of space foods. U.S. Air Force Aerosp. Med. Sch. Contract Rep. F-43-609-68-C-0015.

Labuza, T. P. 1971. Kinetics of lipid oxidation in foods. Crit. Rev. Food Technol. 2:355.

Labuza, T. P. 1979. A theoretical comparison of losses in foods under fluctuating temperature sequences. J. Food Sci. 44:1162.

Labuza, T. P. 1980. The effect of water activity on reaction kinetics of food deterioration. Food Technol. 34(4):36.

Labuza, T. P. 1982. Shelf-Life Dating of Foods. Food and Nutrition Press, Westport, CT.

Labuza, T. P. 1984a. Application of chemical kinetics to deterioration of foods. J. Chem. Educ. 61:348.

Labuza, T. P. 1984b. Moisture Sorption: Practical Aspects of Isotherm Measurement and Use. Am. Assoc. Cereal Chem., St. Paul, MN.

Labuza, T. P. 1985. An integrative approach to food chemistry: Illustrative cases. Pages 913-938 in: Food Chemistry, 2nd ed. O. R. Fennema, ed. Marcel Dekker, New York.

Labuza, T. P., and Contreras-Medellin, R. 1981. Prediction of moisture protection requirements for foods. Cereal Foods World 26:335.

Labuza, T. P., and Kamman, J. F. 1983. Reaction kinetics and accelerated tests: Simulation as a function of temperature. Pages 71-115 in: Computer Aided Techniques in Food Technology. I. Saguy, ed. Marcel Dekker, New York.

Labuza, T. P., and Riboh, D. 1982. Theory and application of Arrhenius kinetics to the prediction of nutrient losses in foods. Food Technol. 36(10):66.

Labuza, T. P., and Schmidl, M. K. 1985. Accelerated shelf-life testing of foods. Food Technol. 39(9):57.

Labuza, T. P., and Schmidl, M. K. 1986. Advances in the control of browning reactions in foods. Pages 65-95 in: The Role of Chemistry in the Quality of Processed Food. O. Fennema, W. H. Chang, and C. Lii, eds. Food and Nutrition Press, Westport, CT.

Labuza, T. P., and Schmidl, M. K. 1988. Use of sensory data in the shelf life testing of foods: Principles and graphical methods for evaluation. Cereal Foods World 33:193.

Labuza, T. P., and Taoukis, P. S. 1990. The relationship between processing and shelf life. In: Foods for the 90s. G. G. Birch, ed. Elsevier, Amsterdam. (In press)

Labuza, T. P., Mizrahi, S., and Karel, M. 1972. Mathematical models for optimization of flexible film packaging of foods for storage. Trans. Am. Soc. Agric. Eng. 15:150.

Lindeman, E. 1953. The change in fat content of sorghum and maize germ during storage. Stule 5:139.

Lingnert, H. 1980. Antioxidative Maillard reaction products: Applications. J. Food Process. Preserv. 4:219.

Lingnert, H. and Waller, G. 1983. Stability of antioxidants formed from histidine and glucose by the Maillard reaction. J. Agric. Food Chem. 3:27.

Megremis, C. 1990. Stabilizing extruded wheat flakes with tocopherols. Cereal Foods World. (In press)

Mizrahi, S., Labuza, T. P., and Karel, M. 1970. Computer aided predictions of food storage stability: Extent of browning in dehydrated cabbage. J. Food Sci. 356:799.

Monahan, E. J. 1988. Packaging of ready-to-eat breakfast cereals. Cereal Foods World 33:215.

Nationwide Food Consumption Survey 1977-1978. 1984. U.S. Dep. Agric. Rep. I-2.

Ragnarsson, J. O., and Labuza, T. P. 1977. Accelerated shelf-life testing for oxidative rancidity in foods—A review. Food Chem. 1:291.

Sakamaki, C., Gray, J. I., and Harte, B. R. 1988. The influence of selected barriers and oxygen absorbers on the stability of oat cereal during storage. J. Packag. Technol. 2:98.

Saltmarch, M., and Labuza, T. P. 1982. Nonenzymatic browning via the Maillard reaction in foods. Diabetes 31(6) Suppl. 3:29.

Salwin, H. 1959. Defining minimum moisture contents for dehydrated foods. Food Technol. 13:594.

Shmigelsky, S., Caldwell, E. F., and Postweiler, J. E. 1964. The effect of several different BHT treatments on post-storage acceptance of a shredded oat cereal. Food Technol. 18:129.

Stuckey, B. N. 1955. Increasing shelf life of cereals with phenolic antioxidants. Food Technol. 9:585.

Taoukis, P. S., El Meskine, A., and Labuza, T. P. 1988. Moisture transfer and shelf life of packaged foods. Pages 243-261 in: Food and Packaging Interactions. J. H. Hotchkiss, ed. Am. Chem. Soc., Washington, DC.

U.S. Code of Federal Regulations. 1988. 21 CFR 137.235, 137.260, 137.305, 137.350.

Zabik, M. E. 1987. Impact of ready-to-eat cereal consumption on nutrient intake. Cereal Foods World 32:234.

Chapter 11

Cereal Nutrition

ROBERT O. NESHEIM
HAINES B. LOCKHART

Long before people learned to cultivate the grasses that are today's cereal grains, they relied upon such grains as a source of nutrients. It is thus only natural that breakfast cereals made from the cereal grains through modern processing techniques have become primary contributors of nutrients to our diets. The nutrients provided by this first meal of the day include those that are indigenous to the cereal grain as well as some that are added in the manufacture of the cereal.

The nutrient contribution of cereals applies to all age groups. Processed cereals are usually the first solid food fed to infants, and the cereal feeding frequently is the first of the day. No one can doubt the popularity of breakfast cereals among children; all one has to do is look on store shelves at the large number of breakfast cereals designed to appeal to children. According to a study by Morgan et al (1981) among children aged 5–12, those who ate ready-to-eat breakfast cereals three or more times a week consumed significantly less fat and cholesterol and more fiber, B-vitamins, and vitamins A and D than those who ate no ready-to-eat cereal at breakfast. Cereals made a contribution to adults as well. Morgan et al (1986) investigated breakfast consumption patterns of adults aged 50 and over and concluded that, for all age and sex classes, consumption of ready-to-eat cereal at breakfast increased the average daily intake of all vitamins and minerals, particularly those identified as underconsumed by elderly individuals. A study made by the U.S. Department of Commerce found that the per capita consumption of cereal breakfast foods is 12–14 lb per year, with ready-to-eat cereals representing approximately 70% of that amount. In a survey made of the Better Homes and Gardens Consumer Panel, 90% of the panel indicated that they kept breakfast cereals in stock in their households.

The effectiveness of the nutritional contributions of a cereal breakfast is attested to by the often-referred-to Iowa Breakfast Studies. Made to assess the dietary importance of breakfast, those studies covered a 10-year period and involved 121 people (Anonymous, 1973). They demonstrated that a nutritious breakfast improved both the physical and mental performance of children and adults. As defined in the Iowa Breakfast Studies, a basic cereal and milk breakfast consists of citrus fruit or juice, 1 oz of cereal with 4 oz of milk, two slices of enriched bread with spread, and 8 oz of milk.

Breakfast Cereals as a Source of Protein

The protein content of cereal grains is small relative to their carbohydrate content. For example, white wheat (the kind used in shredded wheat cereals) contains 75.4% carbohydrate but only 9.4% protein. For wheat made into flour, the ratio is even more in favor of the carbohydrate. With breakfast cereal, the situation is exacerbated when sugar is added to the formula. Because of the high ratio of carbohydrate to protein, the protein content is frequently discounted. The protein of breakfast cereals is often given short shrift because the protein in cereal grains is known to be deficient in certain essential amino acids, such as lysine. However, the combination of cereals with milk helps to alleviate this deficiency in many cases.

Although the protein quality of breakfast cereals as individual foods does not compare favorably with that of animal products like meat and milk, this is not to say that cereal grain proteins in general are as devoid of specific amino acids as gelatin is devoid of tryptophan. Mixtures of cereal grains eaten in large enough quantities will supply all of the essential amino acids needed for growth and maintenance. The cereal grains certainly make their contribution to human protein needs, as attested to by the millions of people throughout several millenia who survived primarily on such grains. Their diets were sometimes supplemented with legumes and only occasionally with sources of animal protein, from meat, fish, and milk. It is reasonable to assume that breakfast cereals can make the same contributions as the cereal grains did, provided the protein is not significantly diminished or damaged by processing.

As might be expected, the protein content of cereal grains and their fractions varies from grain to grain. The protein levels of the cereal grains cover a broad range, as illustrated in Table 1 (Lockhart and Nesheim, 1978). (The protein data in Table 1 are given in grams of nitrogen per 100 g of product. Wheat germ is high in protein content (22.9%), whereas its parent wheat product contains 12.2%. On the other hand a flour of 70–80% extraction from the same or similar wheat contains only 10.9% protein. In one case, fractionation of the original cereal grain increases its protein content; in the other, it results in reduced protein content. Brown rice contains 7.5% protein, but polished rice has 6.7% protein. Corn and oats contain about 9.5% and 15.1% protein, respectively. The protein content of a breakfast cereal reflects, at least in part, the protein content of the cereal grain(s) from which it is made.

The protein contents of 16 representative ready-to-cook (hot) and ready-to-eat breakfast cereals, along with other data taken from their nutritional labeling, are given in Table 2.

The quality of protein of cereal grains as determined by their amino acid contents varies appreciably from grain to grain and among their fractions. Table 1 gives the data for the eight amino acids known to be essential for human nutrition, plus cystine, tyrosine, arginine, and histidine. The high proportion of glutamic acid and proline and the low proportion of lysine, methionine, and tryptophan are characteristic of cereal storage proteins. Lysine appears to be the critical amino acid affecting protein quality among the cereal grains. Note that the lysine level of opaque-2 corn is approximately 50% higher per gram of nitrogen than that of regular corn. This may very well account for the fact that opaque-2 corn has a protein efficiency ratio (PER) of 2.3, versus 1.2 for regular corn. Opaque-2 corn has a higher ratio of germ to endosperm than regular corn, and germ has a high level of lysine. Wheat germ also has a high level of lysine, which probably influences its PER of 2.5, rivaling casein itself. According to Shukla (1975), oat protein, which has a PER of 1.9, is different from other cereal proteins in that 78–80% of the protein is in the form of salt-soluble globulins. Furthermore, the amino acid composition of oat globulin is significantly different from that of other cereal proteins. The remaining 20% or so of the oat protein is composed of not more than 1% albumins, 10–15% alcohol-soluble prolamines, and 2–5% glutelins. The high percentage of globulin of relatively high lysine content is the main reason for the higher overall lysine content of oat protein and for its good nutritive value.

The quality of breakfast cereal protein is obviously related to the quality of the protein of the cereal grains(s) used. Care must be taken, however, that the protein is not damaged in processing. For example, when a reducing sugar such as glucose or fructose is blended with a cereal grain material and the mixture is heat-processed, the resulting Maillard reaction (nonenzymatic browning) makes the lysine physiologically unavailable. This effectively reduces the biological value of the protein. The process of puffing a cereal also reduces its protein

TABLE 1
Protein and Amino Acid Contents of Cereal Grains[a]

Grain	Protein (g/100 g)	Amino Acids[b]												Nitrogen (g/100 g)
		Iso	Leu	Lys	Met	Cs	Phe	Ty	Thr	Try	Val	Ag	Hs	
Barley	11.0	224	417	216	104	142	321	194	207	96[c]	315	295	132	1.88
Maize														
Normal	9.5	230	783	167	120	97	305	239	225	44[d]	303	262	170	1.52
Opaque-2	11.9	194	508	256	84	105	267	236	199	79	298	414	209	1.91
Oats	15.1	236	454	232	105	167	313	206	207	79[c]	319	393	131	2.59
Proso millet	8.1	405	762	189	160	...	308	...	147	49	407	294	119	1.39
Rice														
Brown	7.5	238	514	237	145	67	322	218	244	78[c]	344	516	156	1.26
Polished	6.7	262	514	226	133	96	303	200	207	84[c]	361	473	146	1.13
Rye	11.0	219	385	212	91	119	276	120	209	46[c]	297	286	138	1.89
Sorghum	10.1	245	832	126	87	94	306	167	189	76	313	193	134	1.62
Wheat	12.2	204	417	179	94	159	282	187	183	68[c]	276	288	143	2.09
Germ	22.9	225	433	407	122	130	257	294	265	66[c]	314	513	180	3.95
Gluten	...	258	433	89	100	131	324	229	159	63[d]	266	188	135	...
Bran	13.6	209	415	270	102	168	263	197	223	80[c]	315	490	195	2.16
Flour														
80–90% extr.	11.7	232	379	159	97	127	276	186	192	68	270	259	121	2.05
70–80% extr.	10.9	228	440	130	91	159	304	145	168	67[c]	258	221	130	1.91
60–70% extr.	9.2	217	400	113	87	142	291	132	153	58[c]	240	193	121	1.61

[a]Source: Lockhart and Nesheim (1978); used by permission. Data from FAO (1970) and Pickett (1966).
[b]Values are in mg/g of N, derived by column chromatographic method except as noted.
[c]Derived by microbiological methods.
[d]Derived by chemical methods.

TABLE 2
Nutrition Information on Representative Breakfast Cereals[a,b]

	Quaker			Kellogg			General Mills		Post		Ralston		Quaker		Nabisco	
	Rolled Oats	Farina	Instant Oatmeal	Corn Flakes	Raisin Bran	Fruit Loops	Cheerios	Wheaties	Raisin Bran	Grape Nuts	Cookie Crisp	Wheat Chex	Cap'n Crunch	Life Cereal	Corn Bran	Fruit Wheats
Calories	110	100	90	100	120	110	110	110	80	100	110	100	120	120	120	100
Protein, g	5	3	4	2	3	2	4	3	2	3	1	3	1	5	2	2
Carbohydrates, g	18	22	18	24	31	25	20	23	22	23	25	23	24	19	24	25
Fat, g	2	0	2	0	1	1	2	1	0	1	1	0	2	2	1	0
Dietary fiber, g	*[c]	1	2.8	1	5	1	2	2	4	2	*	2	*	*	5	3
Protein, % U.S. RDA	6	4	6	4	6	2	6	4	4	4	2	4	2	10	2	4
Other nutrients, % U.S. RDA																
Vitamin A	*	*	25	25	25	25	25	25	25	25	*	*	*	*	*	15
Vitamin C	*	*	*	25	*	100	25	25	*	*	*	*	*	*	*	*
Thiamin	10	10	25	50	25	25	25	25	25	25	25	25	25	25	20	25
Riboflavin	*	4	10	50	25	25	25	25	25	25	15	4	15	25	15	25
Niacin	*	6	15	50	25	25	25	25	25	25	25	25	25	25	25	25
Calcium	4	4	15	*	2	*	4	4	*	*	*	*	*	6	2	*
Iron	4	45	25	10	100	25	45	25	25	45	25	25	25	45	45	10
Vitamin D	*	*	*	10	10	10	10	10	10	10	*	*	*	*	*	10
Vitamin B-6	*	*	20	50	25	25	25	25	25	25	25	25	25	25	25	25
Folic acid	*	*	30	50	25	25	*	25	25	25	25	25	25	25	25	25
Vitamin B-12	*	*	*	*	25	*	25	*	25	25	25	25	15	*	15	25
Phosphorus	8	2	10	*	15	2	10	8	10	8	*	8	2	10	2	6
Magnesium	*	*	10	*	15	2	10	8	10	8	*	8	2	*	2	6
Zinc	*	*	4	*	25	25	6	4	10	8	15	4	15	2	15	10
Copper	*	*	6	*	10	2	4	6	8	6	*	6	*	15	*	6
Pantothenic acid	*	*	*	*	*	*	*	*	*	*	20	*	20	*	20	*

[a]Data are taken from labels of these products.
[b]Per 1-oz serving.
[c]Asterisk indicates that the product contains less than 2% U.S. RDA of this nutrient or that no claim is made on the label.

quality. After a grain such as wheat or rice is puffed, its protein quality is so low that it is below the minimum level of quality that allows the product to be claimed as a source of protein in the United States under the nutrition labeling regulations. This is why an amount of protein per serving is not claimed on the labels of those puffed grains.

Although the protein quality of breakfast cereal can be harmed by formulation and processing, it can also be improved. For example, in Life, a ready-to-eat cereal marketed by The Quaker Oats Co., the protein sources are oats, soy flour, and the amino acid lysine. The combination of these protein sources at the levels used results in a protein quality equal that of the animal protein, casein. Because vegetable proteins are not all low in the same amino acids, it is possible to combine vegetable proteins such as oats and soy to give a protein quality approaching that of animal protein. Oat protein is a fair source of lysine and a good source of sulfur amino acids. On the other hand, soy protein is an excellent source of lysine and a poor source of sulfur amino acids. When the two proteins are combined in the proper proportion, the proteins complement each other and the resultant mixture has a protein quality close to that of casein. The small amount of lysine added to the mixture makes the protein equal to casein in quality.

Breakfast Cereals as a Source of Fats and Fatty Acids

Fatty acids occur in appreciable proportions in cereal grains in three principal types—neutral lipid, phospholipid, and glycolipid. Total lipids and individual fatty acids in cereal grains and some of their products are shown in Table 3 (Lockhart and Nesheim, 1978). Table 2 shows the fat content in grams per 1-oz serving of representative breakfast cereal products.

The distribution of individual fatty acids in cereal grains is unique to each grain. The fat of grains has a relatively high content of the essential fatty acid linoleic acid. In corn, wheat, rye, and barley, linoleic acid makes up 57-62% of the total fatty acids in the kernel or in the whole-grain flours of these grains. Saturated fatty acids represent less than 25% of the total for most grains. Palmitic acid is usually the major saturated acid, and oleic, the major monounsaturated acid. The three fatty acids linoleic, palmitic, and oleic together represent about 90–97% of the total fatty acids in most food grains. The content of linolenic acid ranges from 1-2% in corn and rice to 8% in rye. Although oats contain appreciably more total linoleic acid than the other cereal grains, it constitutes only 39% of the total lipid because of the high total lipid content of oats, representing 1.8 times as much as is present

TABLE 3
Fatty Acid Composition of Cereals and Related Products[a,b]

Food	Water	Total Lipid	Fatty Acid[c]									
			Saturated					Unsaturated				
			Sum	14:0	16:0	18:0	20:0	Sum	16:1	18:1	18:2	18:3
Barley, whole grain	14	2.8	0.48	0.01	0.45	0.02	0	1.52	0.01	0.24	1.14	0.13
Corn (*Zea mays*)												
Whole grain, raw	13.8	4.1	0.47	0	0.40	0.08	0.01	3.07	0.01	0.91	2.12	0.03
Flour	12	2.6	0.30	0	0.25	0.04	0.01	2.00	—	0.64	1.34	0.02
Germ	0	30.8	3.93	0	3.30	0.54	0.099	25.57	0.04	7.58	17.7	0.25
Grits, degermed, enriched or unenriched, dry form	12	0.8	0.09	0	0.08	0.01	—[d]	0.60	—	0.18	0.41	0.01
Millet, pearl (*Pennisetum glaucum*)												
Whole grain	11.8	4.1	0.86	0	0.68	0.16	0.02	2.67	0.02	0.83	1.69	0.13
Rolled oats (*Avena sativa*)												
Dry form	8.3	7.4	1.37	0.02	1.21	0.10	0.04	5.65	0.02	2.60	2.87	0.16
Rice (*Oryza sativa*)												
Brown, dry form	12	2.3	0.62	0.03	0.54	0.04	0.01	1.36	0.01	0.54	0.78	0.03
White, fully milled or polished, enriched, dry form	12	0.8	0.21	0.01	0.19	—	—	0.47	—	0.19	0.27	0.01
Rye (*Secale cereale*)												
Whole grain	12.1	2.2	0.27	—	0.25	0.02	0	1.30	0.01	0.22	0.95	0.12
Sorghum (*Sorghum vulgare*)												
Whole grain	11	3.3	0.48	0.01	0.44	0.03	0	2.74	0.04	1.15	1.46	0.09
Wheat (*Triticum aestivum*)												
Whole grain												
Hard red spring	14	2.7	0.37	—	0.36	0.01	—	1.56	0.01	0.25	1.20	0.10
Hard red winter	14	2.5	0.35	—	0.33	0.02	—	1.47	0.01	0.28	1.08	0.10
Soft red winter	14	2.4	0.35	—	0.33	0.02	0	1.40	0.01	0.25	1.07	0.07
White	14	2.0	0.30	—	0.28	0.02	0	1.14	0.01	0.18	0.88	0.07

[a]Source: Lockhart and Nesheim (1978); used by permission. Data from Weihrauch et al (1976).
[b]In grams per 100 g of food, edible portion.
[c]14:0 = myristic, 16:0 = palmitic, 18:0 = stearic, 20:0 = arachidic, 16:1 = palmitoleic, 18:1 = oleic, 18:2 = linoleic, 18:3 = linolenic.
[d]Dashes denote <0.005 g.

in corn or millet and well over three times that in wheat. The high polyunsaturated levels of cereal grain lipids can make an appreciable contribution to essential fatty acid needs and, because of a favorable polyunsaturated-to-saturated (P/S) ratio, can also be of use in diets designed to aid in maintaining normal blood cholesterol levels.

In most grains, fat is not distributed evenly within the kernel but is concentrated primarily in the germ and is lowest in the endosperm. In corn germ, for example, the fat content is seven times the fat content of the whole grain. Oats are an exception to both these rules. According to Shukla (1975), not more than 14–15% of total oat lipids is recoverable from the germ, the balance being associated with the endosperm.

Fat is frequently added to breakfast cereals in their manufacture, and the fat selected may be a liquid, saturated fat such as coconut oil. The addition of saturated fat to cereal grains appreciably alters both the fatty acid distribution of the mixture and the P/S ratio. In the case of the cereal grains themselves, the P/S ratio is in excess of 1.0; however, this is changed when a fat with a high saturated fat level is added to the grain. As an example, a hypothetical breakfast cereal composed of 20% each of whole corn, bran, hard red spring wheat, milled oats, and polished rice, with the balance being nonfat ingredients, would have a P/S ratio of 2.89. If coconut oil, which contains 86.5% saturated fat, most of which is lauric and palmitic acids, were added to the formula at the level of 5%, the P/S ratio would be reduced to 0.30. Although the amount of fat per serving of the breakfast cereal would be increased and the ratio of the fatty acids to one another would change, the *amount* of polyunsaturated fat per serving would be essentially the same for both the cereal with no added fat and the one containing the coconut oil.

Breakfast Cereals as a Source of Carbohydrate

The primary form of carbohydrate in breakfast cereals is starch. The major portion of the carbohydrate in this form in cereal grains is found in the endosperm. On the other hand, water-soluble carbohydrates (chiefly sucrose and raffinose) are found in the wheat embryo and account for the embryo's sweet taste (Aykroyd and Doughty, 1970). Many breakfast cereals contain added carbohydrate in the form of sugar. The starch and the sugar, if present, are the main source of calories in breakfast cereals.

The fraction of carbohydrate that has received considerable attention for the past few years is fiber. Dietary fiber has been defined as that part of plant material taken in the diet that is resistant to digestion by the secretions of the human gastrointestinal tract (Trowell, 1972).

It comprises a heterogeneous group of carbohydrate compounds including cellulose, hemicellulose, pectin, and a noncarbohydrate substance, lignin. These substances, which form the structure of plants, are present in the cell walls of all plant parts, including the leaf, stem, root, and seed. The term *dietary fiber* is not synonymous with *crude fiber*, which is defined as the residue of a foodstuff or a feeding material after treatment with boiling sulfuric acid, sodium hydroxide, water, alcohol, and ether. The crude fiber method substantially underestimates the total amount of indigestible carbohydrate in food.

Although fiber is not a nutrient in a caloric sense, Briggs and Spiller (1978) have proposed that it be considered a nutrient despite its indigestibility, since it has beneficial effects on the body similar to those of other nutrients. Among the physicochemical properties of dietary fiber components that have major impact on the gastrointestinal function are water absorption capacity, cation exchange capacity, organic compound absorption capacity, and gel filtration capacity. Translated into physiologic effects, these factors are consistent with the relief of constipation and the aid in prevention and/or treatment of various abnormalities, including diverticular disease, colon cancer, hemorrhoids, and varicose veins. Alleviation of coronary heart disease has also been attributed to the ingestion of physiologically significant amounts of dietary fiber. However, these relationships are complex, and any one of them may relate to one source of cereal fiber and not to another.

The Life Sciences Research Office of the Federation of American Societies for Experimental Biology has reported on the nutritional significance of dietary fiber (Kimura, 1977). They concluded, among other things, that the type of dietary fiber associated with wheat bran has been shown to reduce gastrointestinal transit time, to modify the fecal composition, to increase stool weight, and to reduce caloric intraluminal pressure. In addition, they concluded that dietary fiber has positive therapeutic value in diverticular disease, atonic constipation, and certain hemorrhoidal conditions. Increased dietary fiber provides bulk, gentle laxation, and ease of elimination.

Like wheat, oats contain some insoluble fiber that exerts a positive effect on the gastrointestinal tract. Unlike wheat, oats contain soluble fiber that has unique physiologic properties. In the 1960s, De Groot et al (1963) reported that consumption of rolled oats reduced serum cholesterol. Since then, De Groot's finding has been confirmed in numerous animal tests. In studies on the effects of rolled oats and of oat bran as a part of a diet for hypercholesterolemic men, the oat products were found to reduce the subjects' serum cholesterol (Gould et al, 1980; Kirby et al, 1981). Also, both rolled oats and oat bran have been found to be effective in reducing the serum cholesterol of healthy adults who are already on a diet with the fat modified to reduce cholesterol (Lipid Research Clinics Program, 1984). The oat products reduced the cholesterol even further. Researchers at King's College London found that 150 g of rolled oats per day fed to subjects on a low-fat diet reduced both total cholesterol and low-density lipoprotein cholesterol, whereas a wheat control diet had no significant effect on cholesterol (Turnbull and Leeds, 1987). University of Wisconsin investigators have found that as little as 4% dietary fiber in a processed oat flour in the ration of laboratory rats significantly lowered cholesterol (Shinnick et al, 1988). Processing increased the proportion of the total fiber that was soluble.

An equally interesting aspect of soluble oat fiber is its effect on blood glucose. High-carbohydrate, high-fiber diets were used successfully to reduce insulin requirements of diabetics (Pederson et al, 1982). Jenkins et al (1981) reported that oatmeal consumption produces a lower blood glucose response than does consumption of other cereals. In a study in which a concentrated soluble oat fiber was fed to normal subjects along with a glucose solution, the blood serum insulin levels were appreciably lower than those of the controls, who received no oat fiber concentrate (Braaten et al, 1988). The authors concluded that the soluble oat fiber was the compound in oats that inhibits postprandial rise in glucose and insulin.

The carbohydrate content of cereal products is frequently determined by difference by subtracting the sum of moisture, protein, fat, ash, and crude fiber percentages from 100. The caloric contribution of carbohydrate is then estimated by multiplying the grams per serving of carbohydrate by a constant, usually 4. Since a minor but significant portion of the carbohydrate is dietary fiber, the caloric value of cereals is probably overestimated. In studies made on the caloric value of cereal grain fractions (Lockhart et al, 1980), the caloric values were determined both experimentally and by calculation. The caloric estimate correlated better with the experimental data when dietary fiber rather than crude fiber was used in calculating the carbohydrate by difference.

Breakfast Cereals as a Source of Vitamins and Minerals

Cereal grains are sources of certain B-vitamins and such minerals as iron, magnesium, and copper. When the grains are converted into breakfast cereals, some of these nutrients are lost (e.g., thiamin). Nutrients that are lost during processing are frequently replaced in the breakfast cereal; however, added nutrients are not limited to lost ones or replacement levels and may include nutrients that are not associated with cereal grains at all, such as vitamins A and C. Table 2 shows the vitamin and mineral content per 1-oz serving of representative breakfast cereal products.

Among various scientific groups that have evaluated the fortification of cereal grain products is the American Association of Cereal Chemists. An article accompanying publication of the Association's policy statement on adding nutrients to cereal grain foods (AACC, 1983) pointed out that cereal grain-based foods had been an appropriate vehicle for adding essential nutrients to the diets of the U.S. population for over 45 years. Cereal grain-based products are good for this purpose because of the significant contribution they make to total caloric and protein intake, the widespread and frequent use of them by virtually all segments of the population, the demonstrated technical feasibility of adding nutrients to these products, and the very low cost of adding these nutrients. Although the AACC policy statement excluded breakfast cereals and did not specify what nutrients should be used for fortification of other cereal grain products or how much should be added, it did endorse the addition of essential nutrients to those products for several purposes. These included use of additives:

a) as a public health measure to minimize or correct a specific nutrient insufficiency known or likely to exist in a significant portion of a population, if the product is consumed in reasonable quantities by that population group; b) to provide nutrients in proportion to the total caloric content of the food; c) to prevent nutritional inferiority in products designed to replace traditional foods; d) to restore nutrients lost in processing; or e) to comply with existing government standards and regulations.

In fortifying a breakfast cereal, the first decision concerns what nutrients to add and at what level they should be added. Numerous regulations and guidelines can be used as a basis for making these decisions. One of the earliest in the United States was the Federal Enrichment Standards, dating to World War II, which covered the addition of a few nutrients to many products. At that time, many of the nutrients now used for fortification were not commercially available. The standards for cereal grain products (Table 4) provided for fortification with thiamin, riboflavin, niacin, and iron, and optionally with calcium and vitamin D for some products. As can be seen from the table, the levels were generally moderate.

In 1974, the National Academy of Science proposed extensive changes in the fortification of cereals (Anonymous, 1974). On the basis of the

TABLE 4
U.S. Federal Enrichment Standards[a]

	Vitamin B-1 (mg/lb)	Vitamin B-2 (mg/lb)	Niacin (mg/lb)	Iron (mg/lb)	Calcium[b] (mg/lb)	Vitamin D[b] (IU/lb)
Bread, rolls, buns	1.8	1.1	15	12.5	600	...
Flour	2.9	1.8	24	20	960	...
Self-rising flour	2.9	1.8	24	20	960	...
Corn grits[c]	2.0–3.0	1.2–1.8	16–24	13–26	500–750	250–1,000
Corn meal[d,e]	2.0–3.0	1.2–1.8	16–24	13–26	500–750	250–1,000
Farina	2.0–2.5	1.2–1.5	16–20	13	500	250
Rice	2.0–4.0	1.2–2.4	16–32	13–26	500–1,000	250–1,000
Macaroni products[f]	4–5	1.7–2.2	27–34	13–16.5	500–625	250–1,000
Macaroni with fortified protein[g]	5	2.2	34	16.5	625	...
Nonfat milk macaroni[f]	4–5	1.7–2.2	27–34	13–16.5	...	...
Vegetable macaroni[f]	4–5	1.7–2.2	27–34	13–16.5	500–625	250–1,000
Noodle products	4–5	1.7–2.2	27–34	13–16.5	500–625	250–1,000
Vegetable noodles	4–5	1.7–2.2	27–34	13–16.5	500–625	250–1,000

[a]Source: Code of Federal Regulations (1988).
[b]Optional.
[c]Includes grits, yellow grits, and quick-cooking grits.
[d]Includes white corn meal, bolted white corn meal, self-rising corn meal, yellow corn meal, bolted yellow corn meal, degerminated yellow corn meal, self-rising yellow corn meal.
[e]Maximum for self-rising corn meals is 1,750 mg.
[f]Includes macaroni, spaghetti, and vermicelli.
[g]Edible protein sources may be added to give a total protein of not less than 20%; protein quality not less than 95% of casein.

nutritional information the Academy had reviewed, it concluded that significant segments of the population risked deficiency in vitamin A, thiamin, riboflavin, niacin, vitamin B-6, folacin, iron, calcium, magnesium, and zinc. Other nutrients were also considered, but their inclusion in a cereal grain fortification program was not recommended at that time. The levels of nutrients recommended are given in Table 5. For comparative purposes, the percent U.S. required daily allowances (RDAs) are given, assuming a 1-oz. serving. The Academy recommended that all products based on the major cereal grains be fortified at the proposed levels when technically feasible.

In the middle of 1974, the U.S. Food and Drug Administration (FDA) proposed a series of nutritional quality guidelines covering, among other things, the fortification of ready-to-eat and ready-to-cook (hot) cereals. The FDA proposal for breakfast cereals is given in Table 6. The number of nutrients the FDA proposed for ready-to-eat cereals was fairly large (11), but three were optional.

More recently, the American Medical Association issued a special report on fortification (1982). The Food and Nutrition Board of the National Academy of Sciences—National Research Council and the Expert Panel on Food Safety and Nutrition of the Institute of Food Technologists reviewed the medical association's statement and concurred with its guidelines. Because breakfast cereals are used primarily as breakfast entrees, the report suggests that they may be formulated to provide, per serving, as much as 25% of the U.S. RDA of the vitamins and minerals common to cereals. In addition, the report indicates that cereals could be considered as carriers for a nutrient when a demonstrated need exists. No apparent justification exists for general fortification of cereals intended for older children and adults beyond 25% of the U.S. RDA for vitamins and minerals.

TABLE 5
Fortification for Cereal Grain Products
as Proposed by the National Research Council's Food and Nutrition Board[a]

Nutrient	Per Pound	Per 100 g	% U.S. RDA
Vitamin A, IU	4,329	954	5.4
Thiamin, mg	2.9	0.64	12.1
Riboflavin, mg	1.8	0.40	6.7
Niacin, mg	24.0	5.29	7.5
Vitamin B-6, mg	2.0	0.44	6.2
Folic acid, mg	0.3	0.07	5.0
Iron, mg	40.0	8.81	13.9
Calcium, mg	900.0	198.20	5.6
Magnesium, mg	200.0	44.10	3.1
Zinc, mg	10.0	2.20	4.2

[a]Source: Anonymous (1974).

The information given above suggests that a broad spectrum of nutrients would be acceptable for fortification. The breakfast cereal industry in general has adopted a moderate position with respect to fortification. Although specific nutrients and levels of nutrients vary from product to product, the general approach has been to provide up to 25% of the U.S. RDA of 10–12 nutrients. Exceptions to this include a breakfast cereal that contains 45% of the U.S. RDA of iron, because it is designed to fit into the U.S. Woman, Infants, and Children (WIC) feeding program, and the limited number of highly fortified products that provide 100% of the U.S. RDA for a number of vitamins and minerals.

In terms of the nutritional labeling that would appear on the package, a typical fortification scheme to provide 25% of the U.S. RDA of the principal vitamins and minerals would be set up as shown in Figure 1.

Vitamin A is currently consumed in amounts below the RDA by certain population groups in the United States and elsewhere. Although cereal is not a natural source of this nutrient, the universal consumption of these foods provides an effective means of delivering this nutrient to the consumer. A portion of the vitamin A intake from cereal plus milk is supplied by the milk (H. D. Hurt and T. E. Milling, personal communication, 1987).

TABLE 6
Guidelines for Breakfast Cereals
as Proposed by the U.S. Food and Drug Administration (% U.S. RDA)

| Nutrient | Ready-to-Eat Cereals | | Ready-to-Cook (Hot) Cereals | |
	Proposed	Existing (Cap'n Crunch)	Proposed	Existing (Quaker Instant Oatmeal)
Protein	10	2	10	6
Vitamin A	25	...	...	20
Thiamin	25	25	25	20
Riboflavin	15	15	15	10
Niacin	25	25	25	15
Calcium	15	...	15	10
Iron	25	25	45	45
Vitamin B-6	25	24	...	20
Folic acid	25	25	...	20
Vitamin B-12	...	15		
Magnesium (optional)	25	2		
Zinc (optional)	25	15		
Pantothenic acid	...	20		

[a]Source: FDA (1974).

An estimated 40% of the U.S. population consumes less than the U.S. RDA of vitamin C. Low intake even for a short time may be of importance because the body does not store vitamin C. The vitamin can be economically added to breakfast cereal.

Thiamin, riboflavin, and iron were the added nutrients permitted by the original Federal Enrichment Standards. They are also indigenous to cereal grains. It naturally follows that they are candidates for fortification of breakfast cereals. The half cup of milk served with breakfast cereals contributes 10% of the riboflavin, which, with the 15% in the cereal, makes up 25% of the U.S. RDA for cereal plus milk. As pointed out earlier, in a product formulated to be used in the WIC program, the iron content of the breakfast cereal is increased to 45% U.S. RDA.

The proper intake of calcium by the general population and specific subgroups is currently a subject of debate within the nutritional and medical community. Also, it is difficult to incorporate calcium into a breakfast cereal without adversely affecting the flavor. For these reasons, calcium is not included in the typical fortification scheme already given. Where possible, it would be desirable to add 10% U.S. RDA of calcium to the cereal, letting milk supply the 15% balance. The addition of calcium to cereals may be less urgent because the milk served as part of the "basic breakfast" will supply appreciable calcium.

	% U.S. RDA in Cereal	% U.S. RDA with Milk
Vitamin A	20	25
Vitamin C	25	25
Thiamin	25	25
Riboflavin	15	25
Niacin	25	25
Calcium	*	15
Iron	25	25
Vitamin D	15	25
Vitamin B-6	25	25
Folic acid	25	25
Vitamin B-12	15	25
Zinc	15	15
Pantothenic acid	25	25
*Contains less than 2% U.S. RDA.		

Figure 1. Typical fortification scheme, shown in label format. Note: If the product is designed for the WIC program, it must be fortified with 45% of the U.S. RDA of iron per serving.

Vitamin D is considered to be important in facilitating the biological utilization of calcium. Fortified milk supplies some of the vitamin D in the cereal-milk mixture.

Cereal grains are natural sources of vitamin B-6, folic acid, pantothenic acid, and zinc. In the typical fortification scheme, the levels of vitamin B-6, pantothenic acid, and folic acid are at 25% of the U.S. RDA. Like calcium salts, zinc compounds have an adverse effect on the flavor of a breakfast cereal if the level is too high. For that reason, the level recommended is 15% U.S. RDA.

Generally, only foods of animal origin are regarded as sources of vitamin B-12. As consumers reduce their intake of animal products and increase the consumption of grain-based foods, there is a danger of reducing the vitamin B-12 intake to the point where it could be inadequate. Accordingly, fortifying foods like breakfast cereals with vitamin B-12 is appropriate, and is easily done. The milk served with breakfast cereal will also supply some vitamin B-12.

Conclusion

Considering the experience that humankind has had with cereal grains for generations, it is not surprising that cereals have become popular and nutritious breakfast entrees for people of all ages. Although the protein content of breakfast cereals relative to carbohydrate content is small, cereals still make worthwhile contributions to the protein portion of the diet. This is especially true if the protein is protected from damage during processing or if the quantity and quality of the protein are amplified by formulation. The fats provided by breakfast cereals are primarily unsaturated unless the fat moiety is modified by the addition of saturated fats. The carbohydrates of breakfast cereals are mostly starch, plus sugar if sugar is added to the formula. A small but physiologically significant portion of the carbohydrate is dietary fiber. Cereal grains are naturally good sources of vitamins and minerals. These are frequently added to breakfast cereals to make the cereals even better sources of nutrients. It is also common to fortify breakfast cereals with nutrients that are not found naturally in cereal grains to round out the nutritional contribution of the cereal and assure the nutritional status of consumers.

References

AACC. 1983. AACC issues policy statement on adding nutrients to cereal grain foods. Cereal Foods World 28:373.

American Medical Association. 1982. Special report. The nutritive quality of processed foods: General policies for nutrient additions. Nutr. Rev. 40:95.

Aykroyd, W. R., and Doughty, J. 1970. Wheat in Human Nutrition. Food and Agriculture Organization of the United Nations, Rome.

Anonymous. 1973. Iowa Breakfast Research Studies. Cereal Institute, Chicago, IL.

Anonymous. 1974. Proposed Fortification Policy for Cereal-Grain Products. National Research Council, National Academy of Sciences, Washington, DC.

Braaten, J., Wood, P., Scott, F., Brule, D., Riedel, D., and Poste, R. 1988. High beta-glucose oat gum, a soluble fiber which lowers glucose and insulin after an oral glucose food: Comparison with guar gum. Fed. Proc. 2:1201.

Briggs, S., and Spiller, G. A. 1978. Dietary fiber, glucose tolerance, and diabetes related; New diet therapy emerges. Food Prod. Dev. 12:81.

Code of Federal Regulations. 1988. Title 21, Chapter 137.

De Groot, A. P., Luyen, R., and Pikaar, N. A. 1963. Cholesterol-lowering effect of rolled oats. Lancet 2:303.

FAO 1970. Amino acid content of foods and biological data on proteins. FAO Nutritional Studies No. 24. Food and Agricultural Organizations of the United Nations, Rome, Italy.

FDA. 1974. Food labeling proposal: Nutritional quality guidelines for fortified ready-to-eat breakfast cereals. Fed. Reg. 39:20898.

Gould, M. R., Anderson, J. W., and O'Mahoney, S. P. 1980. Biofunctional properties of oats. Page 447 in: Cereals for Food and Beverages: Recent Progress in Cereal Chemistry and Technology. G. E. Inglett and L. Munck, eds. Academic Press, New York.

Jenkins, D. J. A., Wolver, T. M. S., Taylor, R. H., Barker, H., Fielden, H., Baldwin, J. M., Bowling, A. C., Newman, H. C., Jenkins, A. L., and Goff, D. V. 1981. A physiological basis for carbohydrate exchanges. Am. J. Clin. Nutr. 34:362.

Kimura, K. K. 1977. The Nutritional Significance of Dietary Fiber. Life Sciences Research Office, Federation of American Societies for Experimental Biology, Bethesda, MD.

Kirby, R. W., Anderson, J. W., Sieling, B., Rees, E. D., Chen, W. J. L., Miller, R. E., and Kay, R. M. 1981. Oat-bran intake selectively lowers serum low-density lipoprotein cholesterol concentrations of hypercholesterolemic men. Am. J. Clin. Nutr. 34:824.

Lipid Research Clinics Program. 1984. The lipid research clinic's coronary primary prevention trial results. J. Am. Med. Assoc. 251:351.

Lockhart, H. B., Hensley, G. W., O'Mahoney, J. P., Lees, H. S., and Houlihan, E. J. 1980. Caloric value of fiber-containing cereal fractions and breakfast cereals. J. Food Sci. 45:372.

Lockhart, H. B., and Nesheim, R. O. 1978. Nutritional quality of cereal grains. Page 201 in: Cereals 78: Better Nutrition for the World's Millions. American Association of Cereal Chemists, St. Paul, MN.

Morgan, K. J., Zabik, M. E., and Leveille, G. A. 1981. The role of breakfast in nutrient intake of 5- to 12-year-old children. Am. J. Clin. Nutr. 34:1418.

Morgan, K. J., Zabik, M. E., and Stampley, G. L. 1986. Breakfast consumption patterns of older Americans. J. Nutr. Elderly 5:19.

Pedersen, O., Hjollund, E., Lindskov, H. O., Helms, P., Sorensen, N. S., and Ditzel, J. 1982. Increased insulin receptor binding to monocytes from insulin-dependent diabetic patients after a low-fat, high-starch, high-fiber diet. Diabetes Care 5:284.

Pickett, R. O. 1966. Opaque-2 corn in swine. Page 19 in: Proceedings of the High Lysine Corn Conference. E. T. Mertz and O. E. Nelson, eds. Corn Refiners Association, Washington, DC.

Shinnick, F. L., Longacre, M. J., Ink, S. L., and Marlett, J. A. 1988. Oat fiber: Composition vs. physiological function. J. Nutr. 118:144.

Shukla, T. P. 1975. Chemistry of oats; Protein foods and other industrial products. C.R.C. Crit. Rev. Food Sci. Nutr. 6:383.

Trowell, H. 1972. Fiber: A natural hypocholesterolemic agent. Am. J. Clin. Nutr. 25:464.

Turnbull, W. H., and Leeds, A. R. 1987. Reduction of total and LDL-cholesterol in plasma by rolled oats. J. Clin. Nutr. Gastroenterol. 2:1.

Weihrauch, J. L., Kinsell, J. E., and Watt, B. K. 1976. Comprehensive evaluation of fatty acids in foods. VI. Cereal products. J. Am. Dietet. Assoc. 68:335.

Chapter 12

Quality Assurance in Breakfast Cereal Plants

JOHN E. STAUFFER
ELWOOD F. CALDWELL

"Quality can be thought of as the 'degree or grade of excellence.'. . ." The problem with this definition is that different people look for different things. Even so, there is some consensus among breakfast cereal consumers as to what constitutes a high-quality product. All consumers expect a safe and wholesome product, and they are greatly interested in nutrition. They also look for convenience, acceptable taste, and value. Taken together, these characteristics largely determine the quality of a breakfast cereal and act as useful yardsticks for comparing one product with another.

Quality Assurance Versus Quality Control

Too often the terms *quality assurance* and *quality control* are used interchangeably, with the result that the difference between them has become blurred. *Quality assurance* is a strategic management function that establishes policies related to product design criteria and adopts programs to assure that product characteristics meet these design criteria. *Quality control* is a tactical function that implements the programs identified by quality assurance as necessary to attain the quality goals (Stauffer, 1988).

Quality assurance covers a wide range of programs. It has oversight responsibility in the areas of product planning, manufacturing, customer service, and distribution. Its duties include the approval of specifications for raw materials, additives, processing aids, finished products, labeling, and packaging. Some of these assignments, such as packaging, are quite broad in scope and therefore are treated separately in this book.

319

Nothing in quality assurance can be left to chance. A planned and systematic approach, including the creation of systems to link the various control functions of subsystems, is essential to the effective performance of the quality assurance function. To accomplish this task, the food technologist has available a number of tried and proven methods.

Key Elements of Quality Assurance

RAW MATERIAL SPECIFICATIONS

All breakfast cereals contain as their basic ingredient one or more cereal grains, the most popular ones being wheat, corn, oats, and rice. Caloric sweeteners, such as sugar, corn syrup, molasses, honey, or brown sugar, are frequently added to this base. Noncaloric or low-calorie sweeteners such as aspartame have also begun to find use in ready-to-eat cereals. In addition, flavoring materials such as salt, malt syrup, yeast, cinnamon, cocoa, and flavor enhancers might be used. The cereal may be processed with vegetable oils to enhance texture. Various protein supplements may be included, the most common being nonfat dry milk, whey, caseinate, and soy protein products. Most cereals are fortified with a mix of vitamins and minerals. The golden brown color characteristic of many cereals is primarily due to the Maillard or nonenzymatic browning reaction between reducing sugars and free amino groups from the proteins present. Where the formulation does not otherwise provide a desirable color, artificial color may be added from among annatto extract, beet powder, caramel color, FD&C Yellow Nos. 5 and 6, FD&C Red 40, and FD&C Blue No. 1. Antioxidant preservatives such as butylated hydroxyanisole (BHA) and butylated hydroxytoluene (BHT) are frequently incorporated into cereals or their packaging materials to provide shelf stability against the development of staleness and rancidity. Finally, a growing trend is to blend into cereals such additional characterizing ingredients as dried fruits, nuts, and marshmallow bits (Andres, 1984; Havighorst, 1985; Anonymous, 1986).

The U.S. Department of Agriculture (USDA) has established official standards and grades for important food crops. As an illustration, the grades for corn, oats, wheat, and rice are given in Appendix 1 following this chapter. The USDA has also set specifications for raisins, nuts, and other ingredients of interest to cereal chemists. USDA grades are purely voluntary; however, they form the basis for contracts between suppliers and buyers and are valuable references even when other specifications are used.

Specifications for nonagricultural commodities are available from various sources. Many food ingredients with known chemical identities are included in *Food Chemical Codex* (FCC); these include ascorbic acid (vitamin C), BHA, BHT, L-lysine monohydrochloride, and sodium chloride (salt). FCC specifications are also recognized by the U.S. Food and Drug Administration (FDA) and therefore must be met for all food applications in the United States. The purity of other food additives, such as food dyes, may be specified in statutes or regulations covering the particular ingredients.

In cases where no official standards apply, the supplier may well be the best source of information. Sugar, liquid sucrose, corn syrups, malt flavoring, and caseinate are examples of products with specifications that may vary considerably. However, the supplier should be willing to guarantee in writing that any such product is "food grade" and is in compliance with the food laws. Processing aids, such as an emulsifier used to prepare a premix of vitamins and minerals, although not declared on the cereal label, must nevertheless conform to standard specifications where they exist and must be "food grade."

TEST METHODS

Specifications are meaningless unless the test methods used to determine the attributes are clearly defined. These test methods should be incorporated into a specification by reference, thereby becoming an integral part of it.

Over the years a body of standard analytical procedures has been developed for food products. The most prominent authorities on procedures are the Association of Official Analytical Chemists and the American Association of Cereal Chemists (AACC). Another source, for those monographs included in it, is the FCC.

In 1922 AACC released its first publication, containing eight "approved methods" and five "tentative methods." In 1928 the original 12-page pamphlet was revised and issued as a 176-page hardcover volume. Now in its eighth edition, *Approved Methods of the American Association of Cereal Chemists* (AACC, 1983) comes in a loose-leaf format so that it can be easily updated. It has grown to two volumes totaling more than 1,200 pages and contains many of the latest instrumental techniques (Christensen, 1983; Tarleton, 1984).

New test methods are constantly being developed to provide greater speed, accuracy, convenience, or versatility. They frequently are adaptable to on-line control functions in operating processes (see section on critical control points later in this chapter). Although these advanced procedures are extremely useful, they nevertheless must be calibrated and checked against approved traditional analytical methods. In due course, as experience is gained with new techniques and as they become generally accepted, many of them will be assimilated into the body of standard references.

REGULATORY REQUIREMENTS

Cereal manufacturers must have a thorough knowledge of government regulations that establish specifications for and provide guidance in the proper use of raw materials and supplies. Government regulations specify minimum requirements only and still allow manufacturers wide latitude to adjust recipes to meet the perceived wants and needs of the consumer.

Food Additives and GRAS Substances

Food products are regulated in the United States under the Federal Food, Drug, and Cosmetic Act. A key feature of this Act is the Food Additives Amendment of 1958, which for the first time placed the responsibility on the food manufacturer to demonstrate that all food ingredients are safe. Thus, before any new food ingredient may be used in food, it must be thoroughly tested by industry and an application must be submitted to the FDA.

The Amendment recognized, however, that many foodstuffs, including agricultural commodities, have been eaten for centuries with no deleterious effects. These foods were termed "generally recognized as safe" (GRAS) and were approved under a separate clause. Although there are special exceptions, food ingredients generally fall into one of two somewhat arbitrary classifications for regulatory purposes: "food additives," which require regulatory approval in advance of use, and GRAS substances, which are accepted because their safety is generally recognized as a result of common use without perceptible harm. Because it may not be perfectly clear whether a substance is GRAS, the FDA will consider, upon request, whether or not an ingredient meets the "GRAS test" and, if it does, will affirm this via official publication in the *Federal Register*. Hundreds of substances have had their GRAS status affirmed in this way, and new substances are added to the list from time to time.

Under the law, commodities such as grains, fruits, nuts, salt, and sugar have been accepted as GRAS without formal proceedings. Other breakfast cereal ingredients, such as vitamins, minerals, cinnamon, and cacao, have been recognized as GRAS after going through the review process and are listed in the regulations under 21 CFR 182. However, even food ingredients included in the GRAS list must be "used for the purpose indicated" and "in accordance with good manufacturing practice." The latter requirement includes the provision that the quantity of an ingredient used should not exceed the amount needed to accomplish its intended effect.

Inclusion in the GRAS list does not automatically exclude an ingredient from further scrutiny. If new evidence on food safety comes to light, a GRAS substance may be restricted in its application or may be banned from further use. The proposed development of new strains of cereal grains through genetic engineering is of more than academic interest. Although such products of biotechnology could have enhanced attributes—for instance, greater yields, higher protein, improved pest resistance, or greater drought stamina, from a food safety point of view they could be subject to critical review and to reevaluation of their GRAS status.

Food additives, on the other hand, are listed in 21 CFR 172 and typically include such ingredients as BHA and BHT. The total quantity of both of these additives in dry breakfast cereals cannot exceed 50 ppm. These antioxidants may also be incorporated into packaging materials, provided migration to the cereal is limited to 50 ppm (21 CFR 181.24). Since 1981 the low-calorie artificial sweetener aspartame has been approved for ready-to-eat breakfast cereals as a food additive.

Unavoidable Contaminants

Acknowledging that, with present technology, food cannot be grown and stored entirely free from undesirable contaminants, the regulations make allowances for such impurities. However, it is incumbent upon the food grower and the processor to take every reasonable precaution to minimize the amount of contamination. To limit the levels of such impurities, the government has established tolerances, action levels, or defect action levels (DALs). Any food containing an impurity at or above the prescribed level is considered adulterated and therefore unfit for human consumption. The enforcement of these regulations strictly forbids blending unacceptable product with conforming material in order to bring it into compliance.

As explained more fully in the next section dealing with hazards, the sources of impurities are manifold. They range from residues of the pesticides used on crops to pervasive contamination of the environment by industrial chemicals. Also included are certain microbiological pathogens and remnants of insects and other pests as well as other filth. Established tolerances for pesticides are listed in

21 CFR 193. Table 1 summarizes action levels that pertain to cereal grains, and Table 2 presents the DALs that specifically apply to wheat and corn. Data are also available for nuts and fruits. The government can revise, revoke, or expand any of these standards at any time with due process.

Hazard Analysis and Critical Control Points

By identifying the hazards associated with a food process, control points critical to producing a safe and wholesome product can be determined (Bauman, 1987). Because the production of breakfast cereals has become so complex, proper control over key variables is a necessity.

TABLE 1
Action Levels for Poisonous or Deleterious Substances in Cereal Grains[a]

Contaminant	Action Level
Aflatoxin	20 ppb
Aldrin and dieldrin	0.02 ppm
Benzene hexachloride (BHC)	0.05 ppm
Crotalaria seeds	Average of one whole seed per pound
DDT, DDE, TDE	0.5 ppm
Dimethylnitrosamine[b]	10 ppb
Ethylene dibromide (EDB)[c]	30 ppb
Lindane	0.1 ppm
Mercury[d]	1 ppm

[a]Source: FDA (1987).
[b]In barley malt.
[c]In grain products requiring no cooking before consumption.
[d]In wheat (pink kernels only).

TABLE 2
Defect Action Levels for Corn Meal and Wheat[a]

Product	Defect	Action Level
Corn meal	Insects	Average of one or more whole insects (or equivalent) per 50 g
	Insect filth	Average of 25 or more insect fragments per 25 g
	Rodent filth	Average of one or more rodent hairs per 25 g or average of one or more rodent excreta fragments per 50 g
Wheat	Insect damage	Average of 32 or more insect-damaged kernels per 100 g
	Rodent filth	Average of 9 mg or more of rodent excreta pellets and/or pellet fragments per kilogram

[a]Source: FDA (1984).

Even slight deviations from normal procedures can result in gross inefficiencies and compromised standards.

HAZARDS OF BREAKFAST CEREALS

Chemical contamination from pesticide residues and environmental sources is a constant worry for consumers of agricultural commodities. Many potential contaminants have been identified, but others are lurking in the environment undetected. Accordingly, manufacturers of breakfast cereals must be constantly vigilant for unforeseen chemical hazards as well as for well-publicized contaminants.

Extraneous matter can include tramp metal, stones, wooden sticks, stems and stalks, glass fragments, and peeling paint. With the use of modern high-speed agricultural machinery, some extraneous matter is inevitably swept up with the harvested crop. Other foreign matter can be introduced within the plant. The risk of introducing foreign objects through employee carelessness or poor maintenance of equipment is significant.

Filth is a catchall term for such offensive matter as bird droppings, rodent hairs, insect fragments, and urine stains. Cereal grains, fruits, and nuts are susceptible to such contamination inasmuch as these foodstuffs provide excellent nourishment to pests as well as humans.

Molds are a significant hazard both on cereal grains and in the finished products. Warm, moist conditions favor the growth of these micro-organisms. Aflatoxin, a strong carcinogen discovered in the early 1960s, is produced under certain conditions by a mold (*Aspergillus flavus*) that grows naturally in the field and during harvesting and storage of crops. In addition to the potential problem of raw material contamination, dry breakfast cereals are susceptible to ordinary mold growth if they are exposed to humidity or otherwise pick up moisture during or following processing and packaging.

Breakfast cereals can spoil in a number of ways. Those containing fats or oils, either naturally from the grain or as added from other sources, are subject to rancidity. Fortified products can lose their vitamin potency over time.

Regulatory hazards are another threat. Under the food laws, a product is said to be "misbranded" and therefore prohibited if, among other things, its package is short in weight or its labeling is false in any manner. For example, a product that does not contain the declared amounts of vitamins would not be in compliance.

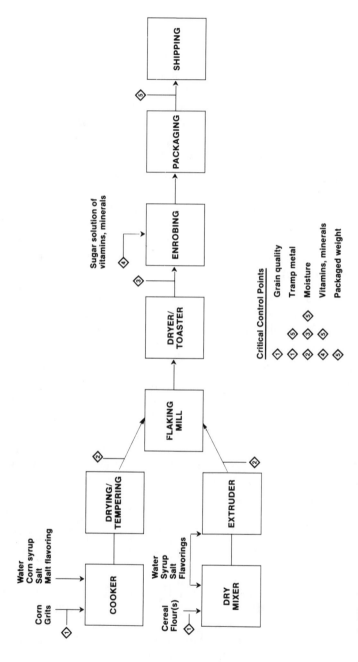

Figure 1. Block diagram of process flows for cereal flakes. Particles going to the flaking rolls consist of either moistened corn grits cooked with malt, sugar, and salt (top) or pellets extruded from a dough made from cereal flour(s) mixed with water, sugar, salt, and other flavors and (in some cases) colors. Numerals indicate critical control points.

CRITICAL CONTROL POINTS

If the hazards of manufacturing a breakfast cereal are known, the critical control points can be determined in a rather straightforward manner. For purposes of instruction, we will look at two ready-to-eat cereal process flows for making cereal flakes (one from corn grits and one from an extruded dough) and one for making a ready-to-cook hot cereal. These processes were chosen to serve as models for discussion only and are not necessarily indicative of the manufacture of particular products.

Figure 1 is a block diagram of the cereal flake processes. Recognizing that whatever is introduced into the process ends up in the finished product, critical control point 1 is established to screen for filth, chemical contaminants, extraneous matter, mold, and offgrade corn grits or cereal flours, as well as for conformance to specification in other functional attributes such as particle size. Because tramp metal, including splinters and filings, may contaminate the product anywhere in the process, critical control point 5 includes a metal detector to scan each sealed carton of cereal. On the same conveyor belt, the cartons pass over a checkweigher that determines and records the packaged weights.

Critical control points 4 and 5 monitor the fortification of the cereal with vitamins and minerals. Even though the vitamins and minerals may be certified by the supplier, spot checks of vitamin potency are made at point 4 with suitable instrumentation, such as the Autoturb microbiological system supplied by Eli Lilly & Co. (Anonymous, 1985). To ensure that the levels of fortification are correct and that the micronutrients are uniformly blended into the product, the finished product is analyzed for an indicator nutrient at point 5.

The control of texture and moisture in the finished cereal flake begins after cooking-drying-tempering or after extrusion, as indicated by point 2 ahead of the flaking rolls. If moisture at that point is too high, the flakes tend to stick to the roll knives and become wrinkled rather than flat. If this happens, they will not blister correctly or at all, and they become hard and flinty rather than crisp and tender. If the tempered grits ahead of flaking are too dry, they will not be gripped and drawn through the "nip," or gap between the rolls. If this happens, no flaking will take place.

The control of moisture in the final product is also critical to maintaining its integrity. Product moisture in excess of 3% when packaged accelerates vitamin C degradation and impairs texture, whereas moisture below 1% may result in off-flavors as well as promote product breakage and rancidity, in either case reducing the shelf life.

Accordingly, moisture is monitored at critical control points 2, 3, and 5. At points 2 and 3, a rapid method of analysis such as near-infrared spectroscopy (Przybyla, 1986) is used to obtain quick determinations of moisture content before flaking and of the flakes as they leave the dryer-toaster oven. The first determination relates to flake texture, as already explained. The second determines whether the key parameters of the oven need to be adjusted to compensate

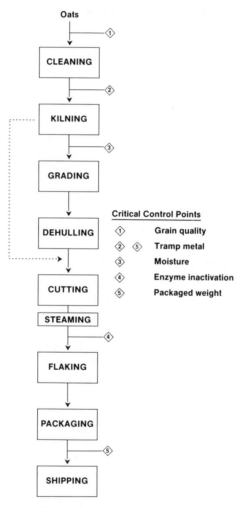

Figure 2. Block diagram of process flow for quick rolled oats. Numerals indicate critical control points. The kilning step, which involves indirect steam heating at atmospheric pressure with or without the addition of direct or sparging steam, more frequently follows rather than precedes dehulling, as indicated by the broken line.

for moisture variations. The oven controls include the air temperature and humidity, air circulation velocity, and retention time of the flakes (Kinney, 1985). The final check of product moisture at point 5 uses the capacitance-dielectric constant principle (Kelly and Swientek, 1986) or other rapid methods.

Critical control points for ready-to-cook rolled oats are shown in Figure 2. Again, critical control point 1 screens for grain quality. Control points 2 and 5 scan for tramp metal and, if necessary, other foreign objects. At point 5, packaged weight is also checked. Point 3 is set up just after the drying step to check product moisture for many of the same reasons that the moisture of the ready-to-eat cereal flakes is controlled, although with different limits appropriate to the product. At point 4, samples are checked to determine whether the lipase enzymes have been inactivated. Because oats are high in natural lipids, they are susceptible to hydrolytic rancidity unless the enzymes have been destroyed.

Corn flakes, first made by W. K. Kellogg in the 19th century, are the most traditional ready-to-eat cereal product known, since it was from them that the industry developed. As discussed in Chapter 2, and as illustrated in the upper flow in Figure 1, most corn flakes are made by cooking so-called "flaking grits" made from degerminated yellow dent corn specifically for flaking. Such grits are relatively uniform (95–100% pass through a 3.5-mesh screen, and about 85% remain on 4- and 5-mesh screens). After the grits are cooked with malt, sugar, and salt, they are cooled and tempered to a uniform moisture and plasticity, flaked on large, smooth rolls operating at differential speeds, and conveyed to the dryer-toaster oven. The moisture content of the tempered grits is clearly a critical control point.

Extrusion, illustrated in the lower flow in Figure 1, has come into use as an alternative way to provide a plastic cereal product for flaking. As described in the Chapter 3 section on continuous cooking, the extruder operates as a high-temperature, short-time cooker, compressing into one operation several steps used in the traditional manufacture of ready-to-eat breakfast cereals. Its primary function is to gelatinize the starch in the grain, but it can also incorporate and/or develop flavors, deactivate enzymes, destroy microorganisms, and denature proteins. Because of the relatively short residence time, nutrient degradation is kept to a minimum. The extruder can produce pellets with given molecular properties, fixed moisture content, and desired density for further processing (Midden, 1989.

Although single-screw extruders were first used as an alternative to batch processing, twin-screw units (typically with corotating screws) have been increasingly used during the 1980s. With either type, however, minor upsets in feed composition or other variables can raise havoc with product quality. For example, a 2% jump in the fat content of the corn has been shown to decrease density 40%, and a 3% rise in moisture increases density 60% (Wiedmann and Strecker, 1987). Start-up and shutdown of the extruder are equivalent to major dislocations in the process. Because biopolymers with low water content can dry up and burn within a few seconds, too rapid a start-up can scorch the product and block the extruder die.

To meet product specifications, tight control must be maintained over three parameters of the extruder-cooker: 1) processing temperature (i.e., temperature of the product within the extruder), 2) residence time, and 3) mechanical shear, as measured by the specific mechanical energy input. Each of these can be independently controlled, although extruder designs differ in the means provided for doing so. The processing temperature can be adjusted by cooling or heating the barrel of the extruder, although the response is sluggish and may present difficulties in scaling a process up to larger capacities. Some extruders provide additional temperature control by means of a steam vent about three-quarters of the way down the barrel. The resulting evaporative cooling is rapid and easy to adjust.

In some extruders, residence time is controlled by varying the pressure in front of the die by means of a throttle valve. This in turn affects the screw fill and thus the residence time.

Perhaps the most effective control loop is the one that determines specific mechanical energy input. This is measured by calculating the quotient of the drive power and the total throughput of all components. It can be controlled by adjusting the water feed, which affects the viscosity of the cereal mash and thereby the mechanical energy that is dissipated.

All three control loops—for energy input, temperature, and residence time—can be monitored by a microprocessor, which is programmed for automatic start-up and shutdown as well as for preset operating conditions (Mulvaney and Hsieh, 1988). Finally, successful operation depends on the accurate feed of corn flour or meal to the extruder. Suitable equipment based on gravimetric feeding is available for this application (K-Tron Corp., 1988).

The color of the flakes emerging from the oven dryer-toaster is not only a key final product characteristic that is influenced by the moisture control points already discussed, but also may serve as a direct process control parameter. Color may be held within a predetermined narrow range of intensity and hue by means of continuous in-line color measurement that feeds back in a closed loop to make oven adjustments (McFarlane, 1988).

Statistical Quality Control

Quality control of modern industrial processes is unthinkable without the use of statistics to interpret and evaluate the significance of large quantities of data. With the help of statistics, such data can be analyzed and converted into simple instructions for the control of the process.

One of the most useful statistical tools is the control chart. Some critical property of a product (e.g., the moisture of a breakfast cereal) is plotted on the vertical axis against a time function (e.g., lot number) on the horizontal axis. When the process is in tight control, the values for moisture show little variation. Should the process be upset, however, the result is immediately evident on the graph as a sharp break in the plotted data (Western Electric Co., 1956). Upper and lower control limits can be established, based on statistical theory, and these lines can be indicated on the control chart. Should data fall outside these limits, corrective action is indicated. Properties such as moisture, density, and color would have both upper and lower reject levels, whereas the percentage of fines or carbonized particles would have only upper reject levels.

The concept of statistical quality control has been extended to the control of variables within a process. Known as statistical process control, this technique monitors the parameters of interest at critical points upstream from finished product inspection (Chowdhury, 1986). In this manner, for example, the moisture of cereal flakes entering the dryer-toaster can be followed. Faulty operation becomes immediately apparent, so that steps can be taken to correct the problem before large quantities of nonspecification product are produced.

Good Manufacturing Practice

A set of standards has been promulgated under the food laws to provide for the proper handling and processing of food products. Commonly known as good manufacturing practices (GMPs), these regulations are delineated in 21 CFR 110. According to these regulations, "a food shall be deemed to be adulterated . . . if it has been prepared, packed, or held under insanitary conditions whereby it may have become contaminated or . . . rendered injurious to health," regardless of whether or not it has actually been contaminated with filth or extraneous matter. This principle is the legal basis for the emphasis that must be placed on sanitation and employee supervision.

The GMPs cover all food categories and perforce are general in nature, addressing such matters as personal hygiene, sanitation, design of facilities, and storage conditions. To be useful to the manufacturer of breakfast cereals, GMPs need interpretation, amplification, and commentary. It is not sufficient to proclaim that "all food-contact surfaces . . . shall be cleaned as frequently as necessary to protect against contamination of food." The manufacturer needs to determine when, how, and under what circumstances each piece of equipment must be cleaned.

SANITATION

First and foremost in a sanitation program is elimination of conditions that cause product contamination or degradation. Accordingly, cleaning of buildings and of processing equipment is essential.

Special attention should be given to cleaning the extruder, if one is used. Most designs are sophisticated and eminently suited to sanitary operation. Twin screws intermesh so that they are self-wiping as they rotate in the barrel. Because the film thickness on the food-contact surfaces is governed by the clearances between extruder parts, it is very small, which prevents the buildup of food particles on the surfaces and minimizes holdup of material in the extruder. The retention time is quite uniform for all material, which facilitates process control.

There is virtually no dead space in a twin-screw extruder, except immediately in front of the die. Some models might have as much as 4 oz of material at this point, but it gets swept out of the barrel because of the large pressure gradient. Correct die design facilitates this flow and equalizes the flow distribution if there is more than one die opening.

All food-contact surfaces are generally made of stainless steel. Because of the wear on the screws and the inner walls of the barrel, these parts may be fabricated from a special high-chromium stainless steel alloy that is very hard. Barrels and screws may be assembled from segments, but no gaskets are used when these parts are bolted together. The faces of each part are machined or ground to a smooth finish to achieve perfect fit without the need for sealing compounds. A recommended feature is that all connections and seals for the hot oil lines used for heating be external to the barrel so that there is no danger that oil can inadvertently leak into the product stream.

Different equipment suppliers have different designs for disassembling their extruders for cleaning and inspection. Baker Perkins features a "clamshell" design that comes apart by unbolting the barrel at longitudinal seams and swinging the top and bottom halves open to expose the screws. Clextral mounts its barrel on an I beam that can be rolled away from the screws after four quick-disconnect clamps are removed. To disassemble the Werner & Pfleiderer extruder, the die head is taken off the barrel and the twin screws are removed after they are uncoupled from the drive shafts.

In practice, extrusion equipment is disassembled infrequently. As long as the extruder line is running smoothly, there is no need to shut it down. The main reasons for stopping are to switch formulations, perform maintenance, cut back on inventory, take vacations, and the like. From a safety point of view, the operation could keep going indefinitely. Sanitation is ensured by the extruder design, the extreme processing conditions, and the insensitivity of the ingredients to microbiological hazards. In fact, frequent shutdowns are said to be abusive to the machinery.

At the end of a production run, which could last hundreds of hours, the equipment can be cleaned in several ways. One approach is to flush out the system with rice hulls, which provide abrasive action and are therefore excellent for purging. Another option is to open up the system and clean all parts with steam hoses. Cleaning in place (CIP) with a mild cleaning solution is a third alternative. When frequent changes are made, such as when running batches of different colors, CIP may have certain advantages.

Cleaning the other equipment, such as the dryer-toaster oven, is more straightforward. As a rule, equipment that handles dry solids should be cleaned by dry methods. The preferred technique is to vacuum surfaces with commercial-duty vacuum cleaners. Air hoses and brushes should never be used because they simply scatter the dust. Scheduling depends on individual experience, but one source suggests that an installation handling bulk flour should be cleaned and inspected monthly (Troller, 1983), the purpose being to interrupt the egg-to-adult life cycle of any insects present. A specific and detailed cleaning protocol for each process line should be developed in collaboration with the equipment manufacturer(s). Because of equipment diversity, the protocols will differ in some (perhaps many or all) respects. One such protocol is outlined in Appendix 2.

PEST CONTROL

Birds and rodents can be a nuisance in cereal plants, as in most food plants and warehouses, but insects cause the most consternation. Insects, both crawling and flying, can be subdivided into two categories: those that are vectors of disease, including houseflies and roaches, and stored-grain insects, a large group comprising weevils, moths, borers, beetles, and mites (USDA, 1978).

Several recognized precautions can be taken to limit insect infestation. Foremost among these are storage in sound, modern structures and thorough cleaning of bins, augers, and other ancillary equipment. Low temperature and moisture are also helpful. Mechanical means such as sifting and mixing are old methods that have proved their effectiveness. By themselves, however, these approaches provide only partial relief, and therefore food processors have come to rely on plant fumigation with chemical pesticides.

A turning point in the history of fumigation was reached in 1984, when the risks associated with the fumigant ethylene dibromide (EDB) came to light. This revelation prompted a review of other fumigants, leading to the banning not only of EDB but also of carbon tetrachloride, carbon disulfide, and ethylene dichloride. Removal of these products narrowed the available choices to phosphine (generated from aluminum phosphide or magnesium phosphide), methyl bromide, and chloropicrin for remedial treatment and malathion, synergized pyrethrins, silicon dioxide, diatomaceous earth, chlorpyrifos-methyl, and pirimiphos-methyl (registration pending) for preventive treatments (Flagstad, 1986).

Pest control without resort to liquid fumigants has shaken the grain industry. The limited options now available usually require distasteful sacrifices such as greater capital investment, higher operating costs, and stricter controls (Stauffer, 1986). Among the alternatives remaining, fumigation with gases still appears to be one of the most promising methods of control. The difference between this approach and controlled atmosphere storage (an emerging technology) is not intrinsic. By combining the best features of both systems, even better results seem possible.

A traditional practice that is still effective where plant and climate conditions permit is the "heat out"—in effect, total plant fumigation with hot air. The whole plant, or a major isolatable area of it, is raised to 120°F (49°C) or more for 24–36 hr. Special heating equipment is required, and the effect of this process on solid-state circuitry that may be installed in modern process control or monitoring equipment must be taken into account.

EMPLOYEE SUPERVISION

Human behavior is the most unpredictable element in food processing. Programs to train employees in GMPs and constant supervision of employees are thus essential in a food plant. The development of job descriptions to meet the needs of quality assurance must be a cornerstone of employee relations.

Responsibility for sanitation can either be assigned to production personnel or delegated to a cleanup crew that takes over at the end of a shift or after the day's work is completed. The advantages of using a special crew are that it comes on the job fresh and alert, premium pay for equipment operators is avoided, attention is focused on the special needs of sanitation, and shortcuts and slipshod performance can be averted (Wenger Mfg., 1988). Regardless of who is responsible for cleanup, the importance of the job must be thoroughly understood by all.

Distribution

The distribution pipeline may stretch from a single production facility into the homes of millions of persons across a nation or around the world. Maintaining the quality of a breakfast cereal as it moves through this conduit is an assignment of considerable challenge. The fact that the product is feather-light, fragile, hygroscopic, and susceptible to spoilage and contamination makes the problem that much more difficult.

SHELF LIFE

As mentioned earlier in the discussion of hazards, spoilage can result from the loss of vitamin potency or the development of rancidity. Shelf life can also be limited by mold contamination or loss of crispness. Packaging material specifications and operation of the package line (covered in Chapter 8) can be as critical to quality assurance as any of the process controls discussed here, or more so. To guarantee the consumer a fresh product at the point of purchase, shelf life must be determined under conditions encountered in distribution.

Apart from redesigning the packaging, several measures can be taken to increase the inherent shelf life of a breakfast cereal. In the discussion of critical control points, we stressed the importance of relatively low product moisture. Antioxidants may be added to those cereals that contain oils and fats to retard rancidity (Coulter, 1988). Vitamins are commonly incorporated into coatings applied to flakes or puffed grains (Dougherty, 1988); these coatings help to protect not only the vitamins but also the base ingredients (Bonner et al, 1975). For ready-to-cook breakfast cereals, which must be vitamin-fortified by different means, encapsulation of micronutrients has proved useful. The protective coating on the microspheres withstands short cooking times, is digestible, and releases its nutrients when eaten. Encapsulation enhances the oxidative stability of ascorbic acid, results in superior taste and odor (by masking the strong flavor of thiamin), and prevents color contamination, such as yellowing caused by dispersion of riboflavin.

Shelf life experiments can be conducted by shipping samples of the product into the field and testing them periodically for deterioration. Although these tests give good results, they are time-consuming and costly. One way around this limitation is to use "accelerated testing." This approach depends on the following time-temperature relationship:

$$Q_{10} = \frac{\text{rate at } (T + 10)}{\text{rate at } T} = \frac{\text{shelf life at } T}{\text{shelf life at } (T + 10)} \ ,$$

where Q_{10} is the quality factor, "rate" is the rate of deterioration, and T is the temperature in degrees Celsius (Labuza and Schmidl, 1985). Thus, data can be collected at an elevated temperature at which deterioration occurs more rapidly, and the results can then be corrected to ambient conditions. Sensory data may be used for accelerated testing (Labuza and Schmidl, 1988).

Assumptions concerning environmental conditions can be far from reality and can lead to significant errors in estimating shelf life. For example, the Technical Association of the Pulp and Paper Industry (TAPPI) has selected 73°F (23°C) and 50% relative humidity as standards for designing packaging materials. To illustrate the pitfalls inherent

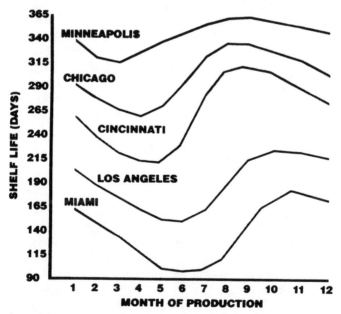

Fig. 3. Shelf life predictions for a ready-to-eat breakfast cereal. (Reprinted, by permission, from Marsh and Wagner, 1985)

in such an assumption, a computer study was undertaken to determine the shelf life for a well-known ready-to-eat breakfast cereal as a function of the month in which it was produced and the location in the United States where it was stored (Marsh and Wagner, 1985). The startling results, shown in Figure 3, indicate wide variations in shelf life depending on the conditions selected.

WAREHOUSING AND SHIPPING

As a breakfast cereal moves through the distribution pipeline from plant to consumer, the number of people who come into contact with

TABLE 3
Common Abuses of Breakfast Cereal During Typical Product Movement[a]

Distribution Point or Stage	Potential Abuses
Producing plant	Packaging line problems
	Product not completely processed.
Loading	
Transportation	Contamination
	Odors
	Materials handling
	Infestation
	Rough transportation
	Temperature abuse
Unloading	
Distribution center	Odors
	Materials handling
	Poor stacking and warehousing
	Poor sanitation
	Poor rotation
	Packaging failure
	Temperature and humidity abuse
Loading	
Transportation	Same as above under "Transportation"
Unloading	
Customer warehouse	Same as above under "Distribution center"
Loading	
Transportation	Same as above under "Transportation"
Unloading	
Retail store	Case cutter cuts
	Temperature and humidity abuse
	Consumer handling
	Poor rotation
	Infestation
	Rough handling
Trip home	
Consumer's home	Temperature and humidity abuse
	Poor rotation or usage
	Rough handling

[a]Modified from Jeffrey (1985).

it multiplies, and opportunities for abuse increase commensurately. The farther a product moves from the manufacturer's control, the less proprietary interest is felt by the people handling it and the less readily available is specific knowledge about the product, such as maximum stacking height and expected shelf life. Special handling of the cereal ceases altogether when it is placed on the shelf with other grocery products (Jeffrey, 1985).

At least 15 different operations can be counted between the manufacturer and the customer. Table 3 shows that the product is vulnerable to abuse at each stage in this chain. Materials handling takes a significant toll. Improper stacking, sloppy warehousing, cuts and punctures from forklift trucks, and rough treatment during transportation add to the problem. Poor rotation of stock, storage with odor-emitting chemicals, and infestation by pests are further dangers.

What precautions can the manufacturer take to control the loss of product quality due to mishandling in distribution? Warehouses and transportation facilities can be inspected on a routine basis. Claims for damaged goods can be investigated. Consumer complaints can be studied and acted upon. Reports from service representatives can be requested and studied. Sales figures can be analyzed for trends, and trade reports of similar activity can be followed. A resourceful manufacturer will find ways to increase the efficiency and reliability of the distribution system.

Consumer Affairs

Consumers do not hesitate to voice their feelings about products. Rather than ignoring or trying to suppress these views, manufacturers to an increasing extent are soliciting comments. Most cereal packages now include a message such as "Guarantee: If you are not satisfied with the quality of this product, return the entire package top for replacement, print your name and address, and tell us why you returned it, where and when you purchased it, and how much you paid for it, and your money will be refunded." Some brands even include a toll-free telephone number to call to express concerns or obtain further information.

Manufacturers mindful of consumer concerns about nutrition, value, and taste are moving to meet these interests. Going a step beyond the legal requirements of nutritional labeling, many brands provide a breakdown of carbohydrates into starch, sugar, and dietary fiber. In an attempt to assuage perceptions of slack fill, a statement is printed on the package to the effect that the product is sold by weight and not volume and that some settling may have occurred during shipping and handling. The consumer is directed to refold the inner liner and close the top after each serving to protect quality and taste. Open-date labeling with the phrase "better if used by" is another step that has been taken to inform the consumer.

Looked upon as consumer education, sales promotion, or simply public relations, these efforts to communicate with the consumer reflect an awareness on the part of the manufacturer of the importance of the end user of the product, who is the ultimate judge of quality.

Product Recall

In the United States, product recall is a voluntary action, not a requirement of the food laws. If a manufacturer is confronted with a problem of nonspecification or possibly harmful product, the decision to institute a recall is the manufacturer's own. This does not mean that the FDA cannot or does not bring pressure to bear on this decision. Through its power to seize products, obtain injunctions, and prosecute criminal acts, the agency has considerable sway over the affairs of food manufacturers.

The FDA has established guidelines to be followed in a product recall, and it will hold the manufacturer accountable for meeting these standards of conduct. A manufacturer faced with the prospect of a product recall is best advised to notify the FDA.

Four classes of product recall, based on the seriousness of the infraction, have been defined by the FDA to convey to the public as clearly as possible the seriousness of any action taken. Class I is specified for the most dangerous cases, such as botulism; class II includes situations that may cause illness but are not life-threatening; class III is reserved for relatively minor violations; and "market withdrawal" is an action by the manufacturer that carries no implication of wrongdoing. Although the classification of a recall does not automatically dictate the depth of the recall, it is the most important consideration. In general, a class I recall shall be made to the consumer level, a class II recall to the retail level, and a class III recall to the wholesale level (product already on supermarket shelves and in the homes of consumers does not need to be returned). An appropriate communication plan must be formulated and implemented to warn the necessary individuals of the action being taken.

References

AACC. 1983. Approved Methods of the American Association of Cereal Chemists, 8th ed. The Association, St. Paul, MN.

Andres, C. 1984. Raisins. Food Process. (Chicago) 45(12):55.

Anonymous. 1985. System measures vitamin potency 80% faster than hand labor. Food Eng. 57(9):113.

Anonymous. 1986. Nuts increase appeal of ready-to-eat cereals. Food Eng. 47(5):37.

Bauman, H. E. 1987. The hazard analysis critical control point concept. Pages 175-179 in: Food Protection Technology. Lewis Publishers, Chelsea, MI.

Bonner, W. A., Gould, M. R., and Milling, T. E. 1975. Ready-to-eat cereal. U.S. patent 3,876,811.

Chowdhury, J. 1986. Quality control moves upstream in CPI plants. Chem. Eng. (N.Y.) 93(8):19.

Christensen, E. A. 1983. AACC revises Approved Methods. Cereal Foods World 28:237.

Code of Federal Regulations. 1983. Title 7, Chapter 810, Part 310.

Code of Federal Regulations. 1988. Title 7, Chapter 810, Parts 404, 1004, and 2204.

Coulter, R. B. 1988. Extending shelf life by using traditional phenolic antioxidants. Cereal Foods World 33:207.

Dougherty, M. E., Jr. 1988. Tocopherols as food antioxidants. Cereal Foods World 33:222.

FDA. 1984. The Food Defect Action Levels. U.S. Food and Drug Administration, Washington, DC.

FDA. 1987. Action Levels for Poisonous or Deleterious Substances in Human Food and Animal Feed. U.S. Food and Drug Administration, Washington, DC.

Flagstad, K. 1986. The "old chemicals" problem. Cereal Foods World 31:799.

Havighorst, C. R. 1985. Those amazing almonds. Food Eng. 57(2):88.

Jeffrey, R. 1985. An overview of product abuse in the food distribution system. Presented at the annual meeting of the American Institute of Chemical Engineers, Chicago, Nov. 10-15.

Johnson, L., Gordon, H. T., and Borenstein, B. 1988. Vitamin and mineral fortification of breakfast cereals. Cereal Foods World 33:278.

Kelly, P. L., and Swientek, R. J. 1986. Compact moisture analyzer provides results in 15 sec. Food Process. (Chicago) 47(11):86.

Kinney, T. B. 1985. How to overcome common dryer control problems. Food Eng. 57(12):106.

K-Tron Corp. 1988. K-Tron LW65 all-digital loss-in-weight system. K-Tron Corporation, Glassboro, NJ.

Labuza, T. P., and Schmidl, M. K. 1985. Accelerated shelf-life testing of foods. Food Technol. 39(9):57.

Labuza, T. P., and Schmidl, M. K. 1988. Use of sensory data in the shelf life testing of foods: Principles and graphical methods for evaluation. Cereal Foods World 33:193.

Marsh, K. S., and Wagner, J. 1985. Computer model looks at the environment to predict shelf life. Food Eng. 57(8):58.

McFarlane, I. 1988. In-line measurement and closed-loop control of the color of breakfast cereals. Cereal Foods World 33:978.

Midden, T. M. 1989. Twin-screw extrusion of corn flakes. Cereal Foods World 34:941.

Mulvaney, S. J., and Hsieh, F.-H. 1988. Process control for extrusion processing. Cereal Foods World 33:971.

Przybyla, A. E. 1986. Analytical instruments simplify procedures. Food Eng. 58(5):142.

Stauffer, C. 1986. Pest control without liquid fumigants. Cereal Foods World 31:807.

Stauffer, J. E. 1988. Preface. In: Quality Assurance of Food. Food and Nutrition Press, Westport, CT.

Tarleton, R. J. 1984. AACC Approved Methods. Cereal Foods World 29:565.

Troller, J. A. 1983. Sanitation in Food Processing. Academic Press, New York.

USDA. 1978. Stored-Grain Insects. U.S. Dep. Agric. Handb. 500. U.S. Department of Agriculture, Washington, DC.

Wenger Mfg. 1988. General Sanitation. Wenger Manufacturing, Inc., Sabetha, KS.

Western Electric Co. 1956. Statistical Quality Control Handbook. Western Electric Co., Indianapolis, IN.

Wiedmann, W., and Strecker, J. 1987. How to automate an extruder. Chilton's Food Eng. Int. 12(3):40.

Appendix 1

Grades for Various Grains

TABLE A-1
Grade Requirements for Corn[a,b]

U.S. Grade	Minimum Test Weight per Bushel (lb)	Broken Corn and Foreign Material (%)	Damaged Kernels	
			Total (%)	Heat-Damaged Kernels (%)
No. 1	56	2.0	3.0	0.1
No. 2	54	3.0	5.0	0.2
No. 3	52	4.0	7.0	0.5
No. 4	49	5.0	10.0	1.0
No. 5	46	7.0	15.0	3.0
Sample Grade	...	...	...	...

U.S. Sample Grade is corn that (1) does not meet the requirements for grades No. 1–5; or (2) in a 1,000-g sample, contains eight or more stones with aggregate weight in excess of 0.20% of the sample weight, two or more pieces of glass, three or more crotalaria seeds (*Crotalaria* spp.), two or more castor beans (*Ricinus communis* L.), four or more particles of an unknown substance(s) or a commonly recognized harmful or toxic substance(s), eight or more cocklebur (*Xanthium* spp.) or similar seeds singly or in combination, or animal filth in excess of 0.20%; or (3) has a musty, sour, or commercially objectionable foreign odor; or (4) is heating or otherwise of distinctly low quality.

[a]Source: 7 CFR 810.404.
[b]For the classes Yellow corn, White corn, and Mixed corn.

TABLE A-2
Grade Requirements for Oats[a]

| U.S. Grade | Minimum Limits | | Maximum Limits | | |
	Test Weight per Bushel (lb)	Sound Oats (%)	Heat-Damaged Kernels (%)	Foreign Material (%)	Wild Oats (%)
No. 1	36.0	97.0	0.1	2.0	2.0
No. 2	33.0	94.0	0.3	3.0	3.0
No. 3[b]	30.0	90.0	1.0	4.0	5.0
No. 4[c]	27.0	80.0	3.0	5.0	10.0
Sample grade					

U.S. Sample grade are oats that:

(1) Do not meet the requirements for the grades U.S. Nos. 1, 2, 3, or 4; or

(2) Contain eight or more stones that have an aggregate weight in excess of 0.2% of the sample weight, two or more pieces of glass, three or more crotalaria seeds (*Crotalaria* spp.), two or more castor beans (*Ricinus communis* L.), four or more particles of an unknown foreign substance(s) or a commonly recognized harmful or toxic substance(s), eight or more cocklebur (*Xanthium* spp.) or similar seeds singly or in combination, 10 or more rodent pellets, bird droppings, or equivalent quantity of other animal filth per 1-1/8 to 1-1/4 qt of oats; or

(3) Have a musty, sour, or commercially objectionable foreign odor (except smut or garlic odor); or

(4) Are heating or otherwise of distinctly low quality.

[a] Source: 7 CFR 810.1004.
[b] Oats that are slightly weathered shall be graded not higher than U.S. No. 3.
[c] Oats that are badly stained or materially weathered shall be graded not higher than U.S. No. 4.

TABLE A-3
Grade Requirements for Wheat[a]

| Grade | Minimum limits of Test Weight per Bushel | | Maximum limits of | | | | | | |
	Hard Red Spring Wheat or White Club Wheat (lb)[b]	All Other Classes and Subclasses (lb)	Heat-Damaged Kernels (%)	Damaged Kernels (Total)[c] (%)	Foreign Material (%)	Shrunken and Broken Kernels (%)	Defects (Total)[d] (%)	Wheat of Other Classes[e] Contrasting Classes (%)	Total[f] (%)
U.S. No. 1	58.0	60.0	0.2	2.0	0.5	3.0	3.0	1.0	3.0
U.S. No. 2	57.0	58.0	0.2	4.0	1.0	5.0	5.0	2.0	5.0
U.S. No. 3	55.0	56.0	0.5	7.0	2.0	8.0	8.0	3.0	10.0
U.S. No. 4	53.0	54.0	1.0	10.0	3.0	12.0	12.0	10.0	10.0
U.S. No. 5	50.0	51.0	3.0	15.0	5.0	20.0	20.0	10.0	10.0

U.S. Sample grade

U.S. Sample grade is wheat that

(1) Does not meet the requirements for the grades U.S. Nos. 1, 2, 3, 4, or 5; or

(2) Contains 32 or more insect-damaged kernels per 100 g of wheat; or

(3) Contains eight or more stones or any number of stones that have an aggregate weight in excess of 0.2% of the sample weight, two or more pieces of glass, three or more crotalaria seeds (*Crotalaria* spp.), two or more castor beans (*Ricinus communis* L.), four or more particles of an unknown foreign substance(s) or a commonly recognized harmful or toxic substance(s), two or more rodent pellets, bird droppings, or equivalent quantity of other animal filth per 1,000 g of wheat; or

(4) Has a musty, sour, or commercially objectionable foreign odor (except smut or garlic odor); or

(5) Is heating or otherwise of distinctly low quality.

[a] Source: 7 CFR 810.2204.

[b] These requirements also apply when Hard Red Spring wheat or White Club wheat predominate in a sample of Mixed wheat.

[c] Includes heat-damaged kernels.

[d] Defects include damaged kernels (total), foreign material, and shrunken and broken kernels. The sum of these three factors may not exceed the limit for defects for each numerical grade.

[e] Unclassed wheat of any grade may contain not more than 10.0 % of wheat of other classes.

[f] Includes contrasting classes.

TABLE A-4
Grades and Grade Requirements for the Classes of Milled Rice[a]

U.S. Grade	Seeds, Heat-Damaged, and Paddy Kernels (Singly or Combined) Total (no. in 500 g)	Heat-Damaged Kernels and Objectionable Seeds (no. in 500 g)	Red Rice and Damaged Kernels (Singly or Combined) (%)	Chalky Kernels[b] In Long-Kernel Rice (%)	Chalky Kernels[b] In Medium- or Short-Kernel Rice (%)	Broken Kernels Total (%)	Broken Kernels Removed by a 5 Plate[c] (%)	Broken Kernels Removed by a 6 Plate[c] (%)	Broken Kernels Removed Through a 6 Sieve[c] (%)	Other Types[d] Whole Kernels (%)	Other Types[d] Whole and Broken Kernels (%)	Color Requirements[b]	Milling Requirements (Minimum)[e]
No. 1	2	1	0.5	1.0	2.0	4.0	0.04	0.1	0.1	…	1.0	Shall be white or creamy	WM
No. 2	4	2	1.5	2.0	4.0	7.0	0.06	0.2	0.2	…	2.0	May be slightly gray	WM
No. 3	7	5	2.5	4.0	6.0	15.0	0.1	0.8	0.5	…	3.0	May be light gray	RWM
No. 4	20	15	4.0	6.0	8.0	25.0	0.4	2.0	0.7	…	5.0	May be gray or slightly rosy	RWM
No. 5	30	25	6.0[e]	10.0	10.0	35.0	0.7	3.0	1.0	10.0	…	May be dark gray or rosy	LM
No. 6	75	75	15.0[f]	15.0	15.0	50.0	1.0	4.0	2.0	10.0	…	May be dark gray or rosy	LM

Sample grade
  U.S. Sample grade shall be milled rice or any of these classes that (1) does not meet the requirements for any of the grades from U.S. No. 1 to No. 6, inclusive; (2) contains more than 15.0% of moisture; (3) is musty, or sour, or heating; (4) has any commercially objectionable foreign odor; (5) contains more than 0.1% of foreign material; (6) contains live or dead weevils or other insects, insect webbing, or insect refuse; or (7) is otherwise of distinctly low quality.

[a]Source: 7 CFR 68.310.
[b]For the special grade Parboiled Milled rice, see 7 CFR 68.315(c).
[c]Plates should be used for southern production rice, and sieves should be used for western production rice; but any device or method that gives equivalent results may be used.
[d]These limits do not apply to the class Mixed Milled rice.
[e]For the special grade Undermilled Milled rice, see 7 CFR 68.315(d). WM = well milled, RWM = reasonably well milled, LM = lightly milled.
[f]Grade U.S. No. 6 shall contain not more than 6.0% of damaged kernels.

Appendix 2

Typical Procedure for Sanitation
of Extruder After Shutdown

If shutdown is for die change or other short time:

1. Unbolt knife drive shaft from knife assembly.
2. Remove knife assembly.
3. Remove extruder die.
   A. Use safety chain to prevent damage to the die.
4. Flush extruder barrel.
   A. Use processing water through inlet head.
   B. Rotate extruder shaft throughout flushing.
   C. Flush until relatively clear water exits extruder barrel.

If shutdown is for die change or for less than 3–4 hr, further disassembly is not required. If shutdown will last overnight or longer, complete disassembly is desirable:

1. Remove downspout between conditioning cylinder and inlet head.
   A. Clean downspout with steam or high-pressure water.
2. Clean knife-bearing assembly.
   A. Hand-clean only with wire brush and scraper.
   B. Do not submerge, as submersion could damage bearing assemblies.
3. Clean extruder die.
   A. Remove product from die orifices with drill rod or other rod of suitable diameter.
   B. Scrape all material from face and back of die plate.
   C. Similarly clean die spacers and adapters used.
4. Remove head jacket plumbing and head clamps.
   A. Remove thermocouples and pressure-sensing senders (where used).
   B. When sliding off and handling heads, take care not to damage quick-connect couplings by dropping or rolling them on hard surfaces. Rough handling may lead to leakage and may restrict the flow of steam and water to head jackets.
   C. Bathe-wash heads. Detergents or caustic soda may be used on heads.
   D. If extruder has a split barrel, remove all clamping bolts and open the barrel.
5. Clean shaft.
   A. Do NOT remove screws from shaft.
   B. Spray-wash shaft with steam or high-pressure water. Detergent or caustic soda may be used.
6. Clean circular bin discharger.
   A. Open clean-out door.
   B. Wash out inside with steam or high-pressure water.

7. Clean screw feeder.
   A. Remove top section of feeder housing.
   B. Wash out with steam or high-pressure water.
8. Clean mixing cylinder.
   A. Fabricate flange plate for outlet of mixing cylinder with 3-in. valve installed. Bolt or clamp to outlet flange.
   B. Fill mixing cylinder half full with warm water, add detergent, and run mixing cylinder 30–45 min.
9. Clean frame, shaft, and screws.
   A. Close watertight door on control panel.
   B. Wash frame, motor, and shaft.
   C. Steam, high-pressure water, detergents, or caustic soda solutions may be used.
10. Clean dryer/cooler.
   A. Because of possible bacterial growth, under normal circumstances do NOT wash dryers with steam, water, or cleaning solutions.
   B. Remove excess moisture that may provide an incubator for salmonellae and other bacterial agents.
   C. Remove fines with positive air or vacuum system.
   D. Remove burned-on or sticky product by scraping with wire brush over conveyor trays.
   E. If for some reason dryer must be washed, after wetting, the dryer should be set at about 350°F (177°C) and allowed to run for 1.5–2 hr.

Source: Wenger Manufacturing, Inc., Sabetha, KS. Used by permission.

Appendix A

Partial List of Manufacturers of Equipment for Processing and Packaging of Breakfast Cereals

The editors have assembled this list as a service to readers. It includes many of the principal manufacturers of specialized items of equipment for processing and packaging breakfast cereals, primarily those located in the United States or international companies actively marketing in the United States. Listings are alphabetically by chapter. No recommendation is implied by the inclusion or omission of any listing. Suppliers of ingredients and package material and of more general equipment or instrumentation are not included.

Chapter 3. Unit Operations and Equipment. I: Blending and Cooking

Blending and Flavoring
Paul O. Abbe, Inc., Little Falls, NJ 07424
Acrison, Inc., Moonachie, NJ 07074
Bepex Corp., Santa Rosa, CA 95402
Buhler, Inc., Minneapolis, MN 55441
Day Mixing Co., Cincinnati, OH 45212
Demaco/DeFrancisci Machine Corp., Ridgewood, NY 11385
Groen Process Equipment Group, Elk Grove Village, IL 60007
Hamilton Kettles Div., Cincinnati, OH 45241
S. Howes Co., Inc., Silver Creek, NY 14136
Lee Industries, Inc., Phillipsburg, PA 16866
Littleford Bros., Inc., Florence, KY 41042
Mixing Equipment Co., Rochester, NY 14603
Spray Dynamics Division, Par-Way Manufacturing Co., Costa Mesa, CA 92627
Sprout Bauer Co., Muncy, PA 17756
Readco/Teledyne, York, PA 17405
Vibra Screw Inc., Totowa, NJ 07512
Wenger Manufacturing, Inc., Sabetha, KS 66534
The Young Industries, Inc., Muncy, PA 17756

Cooking, batch

APV Baker, Inc., Grand Rapids, MI 49504
Lauhoff Corp., Detroit, MI 48207

Cooking, continuous

Anderson International, Cleveland, OH 44105
APV Baker Inc., Grand Rapids, MI 49504
The Bonnot Co., Kent, OH 44240
Buhler Inc., Minneapolis, MN 55441
Buss-Condux, Inc., Elk Grove Village, IL 60007
Cincinnati Milacron, Batavia, OH 45103
Clextral, Inc., Odessa, FL 33556
Mapimpianti, York, PA 17402
Readco/Teledyne, York PA 17405
Sprout Bauer, Inc., Muncy, PA 17756
Textruder Engineering AG/Raymond Automation Co., Norwalk, CT 06851
Wenger Manufacturing Inc., Sabetha, KS 66534
Werner & Pfleiderer Corp., Ramsey, NJ 07446

Lump breaking

Buhler, Inc., Minneapolis, MN 55441
Champion Products, Inc., Eden Prairie, MN 55344
The Fitzpatrick Co., Elmhurst, IL 60126
Franklin Miller, West Orange, NJ 07052
Gruendler Crusher Div./Lukens General Ind., St. Louis, MO 63114
Jacobson Machine Works, Inc., Minneapolis, MN 55427
Jersey Stainless, Inc., Berkeley Heights, NJ 07922
Wyssmont Co., Inc., Fort Lee, NJ 07024

Sizing

AZO, Inc., Memphis, TN 38118
Buhler, Inc., Minneapolis, MN 55441
Kason Corp, Linden, NJ 07036
Rotex, Inc., Cincinnati, OH 45223
Simplicity Engineering, Durand, MI 48429
Sweco, Inc., Houston, TX 77220

Chapter 4. Unit Operations and Equipment. II: Drying and Dryers

Aeroglide Corp., Raleigh, NC 27626
APV Baker, Inc., Grand Rapids, MI 49504
Buhler, Inc., Minneapolis, MN 55441
Carmen Industries, Jeffersonville, IN 47130
Carrier Vibrating Equipment, Inc., Louisville, KY 40233
Demaco/DeFrancisci Machine Corp., Ridgewood, NY 11385
Food Engineering Corp., Minneapolis, MN 55441
National Drying Machinery Co., Philadelphia, PA 19133
Proctor & Schwartz, Inc., Horsham, PA 19044
Renneburg Div./Heyl Patterson, Pittsburgh, PA 15230

C. G. Sargents Sons Corp., Westford, MA 01886
Spooner Inc., Williamsville, NY 14221
Wenger Manufacturing, Inc., Sabetha, KS 66534
Wolverine Corp., Merrimac, MA 01860
Wyssmont Co., Inc., Fort Lee, NJ 07024

Chapter 5. Unit Operations and Equipment III: Tempering, Flaking, and Toasting

Tempering

APV Baker, Inc., Grand Rapids, MI 49504
Buhler, Inc., Minneapolis, MN 55441
Food Engineering Corp., Minneapolis, MN 55441
Lauhoff Corp., Detroit, MI 48207
Proctor & Schwartz, Inc., Horsham, PA 19044
Wenger Manufacturing Inc., Sabetha, KS 66534

Flaking

APV Baker, Inc., Grand Rapids, MI 49504
Blount/Ferrel-Ross, Oklahoma City, OK 73100
Buhler, Inc., Minneapolis, MN 55441
Lauhoff Corp., Detroit, MI 48207
Wenger Manufacturing Inc., Sabetha, KS 66534
Wolverine Corp., Merrimac, MA 01860

Toasting

APV Baker, Inc., Grand Rapids, MI 49504
Buhler, Inc., Minneapolis, MN 55441
Proctor & Schwartz Inc., Horsham, PA 19044
Shouldice Bros., Inc., Battle Creek, MI 49015
Spooner Inc., Williamsville, NY 14221
Wenger Manufacturing Inc., Sabetha, KS 66534
Wolverine Corp., Merrimac, MA 01860

Chapter 6. Operations and Equipment. IV: Extrusion

APV Baker, Inc., Grand Rapids, MI 49504
The Bonnot Co., Kent, OH 44240
Buhler, Inc., Minneapolis, MN 55441
Clextral, Inc., Odessa, FL 33556
Demaco/DeFrancisci Machine Corp., Ridgewood, NY 11385
Dorsey-McComb, Inc., Denver, CO 80222
Egan Machinery, Somerville, NJ 08876
Farrel Co., Ansonia, CT 06401
D. Maldari & Sons, Inc., Brooklyn, NY 11215

Mapimpianti, York, PA 17402
Readco/Teledyne, York, PA 17405
Sprout Bauer, Inc., Muncy, PA 17756
Textruder Engineering AG/Raymond Automation Co., Norwalk, CT 06851
Wenger Manufacturing, Inc., Sabetha, KS 66534
Werner & Pfleiderer, Inc., Ramsey, NJ 07446

Chapter 7. Application of Nutritional and Flavor/Sweetening Coatings

APV Baker, Inc., Grand Rapids, MI 49504
Spray Dynamics Division, Par-Way Manufacturing Co, Costa Mesa, CA 92627
Spraying Systems Co., Wheaton, IL 60188
Transitube/Clextral, Inc., Odessa, FL 33556
United Air Specialists, Inc., Cincinnati, OH 45242
Wenger Manufacturing Inc., Sabetha, KS 66534

Chapter 8. Packaging and Packaging Materials

Bemis Packaging Service Machine Co., Minneapolis, MN 55418
Robert Bosch Corp., South Plainfield, NJ 07080
Eagle Group/Package Machinery Co., Oakland, CA 94621
Hayssen Manufacturing Co., Sheboygan, WI 53082
R. A. Jones & Co., Inc., Cincinnati, OH 45201
James River Corp.
   Flexible Packaging Group, Oakland, CA 94612
   Riegel Division, Milford, NJ 08848
H. J. Langen and Sons, Ltd., Mississauga, Ont., Canada L4V 1H3
Pneumatic Scale Corp., Quincy, MA 02171
Salwasser Manufacturing Co., Reedley, CA 93654
Thiele Engineering Co., Minneapolis, MN 55435
Triangle Package Machinery Co., Chicago, IL 60635
Waldorf Corp., St. Paul, MN 55114

Chapter 9. Hot Cereals

Manufacture

APV Baker, Inc., Grand Rapids, MI 49504
Buhler, Inc., Minneapolis, MN 55441
Carter-Day Co., Minneapolis, MN 55432
Kipp Kelly/Farm King Allied, Inc., Winnipeg, Canada R3C 2W5
Lauhoff Corp., Detroit, MI 48207
Patterson-Kelly Co./Div. Harsco Corp., East Stroudsburg, PA 18301
Rotex, Inc., Cincinnati, OH 45223

Packaging

Bemis Co., Inc., Terre Haute, IN 47808
R. A. Jones & Co., Inc., Cincinnati, OH 45201
Paxall Clybourn Machinery, Skokie, IL 60077
Pneumatic Scale Corp., Quincy, MA 02171
Solbern Div., Howden Food Equipment, Fairfield, NJ 07006
Tools & Machinery Builders, Inc., Ironton, MO 63650
Triangle Package Machinery Co., Chicago, IL 60635

Appendix B

Additional References

Anderson, D. M. W., and Andon, S. A. 1988. Water-soluble food gums and their role in product development. Cereal Foods World 33:844.

Anderson, R. H., Maxwell, D. L., Mulley, A. E., and Fritsch, C. W. 1976. Effects of processing and storage on micronutrients in breakfast cereals. Food Technol. 30(5):110.

Ando, M., Minami, J., Ohnish, F., and Sawada, M. 1977. Manufacture of ready-to-eat rice. U.S. patent 4,166,868.

Ando, M., Minami, J., Takata, M., Ohnishi, F., and Kawamoto, S. 1978. Process for producing instant-cooking rice. U.S. patent 4,233,327.

Anonymous 1983. Instant breakfast in a yogurt pot. Milk Ind. 85(8):13.

Anonymous 1983. Some interesting ideas with flavored cereals. Food Eng. 55(5):69.

Anonymous 1984. Cooker extruder for cornflakes. Food Eng. Int. 9(Oct.):93.

Anonymous 1988. Raisin paste: A new wrinkle in food formulation. Prep. Foods 157(7):148.

Aurand, T. J. 1989. Cranberry—The up and coming fruit ingredient. Cereal Foods World 34:407.

Baggerly, P. A. 1974. Cereal process and product. U.S. patent 3,955,000.

Bakar, J., and Hin, Y. S. 1985. High-protein rice-soya breakfast cereal. J. Food Process. Preserv. 8(3/4):163.

Baker, D., and Holden, J. M. 1981. Fiber in breakfast cereals. J. Food Sci. 46:396.

Barham, H. N., Jr., Berham, H. N., and Barham, D. 1978. Method for the adsorption of solids by whole seeds. U.S. patent 4,208,433.

Boczewski, M. P. 1978. Process for the treatment of oats. U.S. patent 4,220,287.

Bonner, W. A., Gould, M. R., and Milling, T. E. 1973. Ready-to-eat cereal. U.S. patent 3,876,811.

Bonnyay, L., and Elsken, J. C. 1978. Corn wet milling system and process for manufacturing starch. U.S. patent 4,207,188.

Buckholz, L. L., Jr. 1988. The role of Maillard technology in flavoring food products. Cereal Foods World 33:547.

Caldwell, E. F. 1987. The *Consumer Reports* nutritional index for rating breakfast cereals: An analysis. Cereal Foods World 32:248.

Cardozo, M. S., and Eitenmiller, R. R. 1988. Total dietary fiber analysis of selected baked and cereal products. Cereal Foods World 33:414.

Carson, G. 1976. Cornflake Crusade. Ayer Publications, Bala Cymwyd, PA.

Chan, J. K. C., and Wypszyk, V. 1988. A forgotten natural dietary fiber: Psyllium mucilloid. Cereal Foods World 33:919.

Chwalek, V. P., and Olson, R. M. 1978. Combined dry-wet milling process for refining corn. U.S. patent 4,181,748.

Cloud, L. L., Kelly, V. J., and Smalligan, W. J. 1975. Junior cereal and process. U.S. patent 3,956,506.

Cole, K. M.,Jr. 1975. Continuous double coating natural cereal. U.S. patent 4,061,790.

Cook, D. A., and Welsh, S. O. 1987. The effect of enriched and fortified grain products on nutrient intake. Cereal Foods World 32:191.

Coulter, R. B. 1988. Extending shelf life by using traditional phenolic antioxidants. Cereal Foods World 33:207.

Cox, D. S. 1977. Enriched wheat macaroni. U.S. patent 4,158,069.

Dahl, M. J. 1975. Apparatus for continuous puffing. U.S. patent 3,971,303.

Daniels, R. 1970. Modern Breakfast Cereal Processes. Noyes Data Publ., Park Ridge, NJ.

Daniels, R. 1974. Breakfast Cereal Technology. Noyes Data Publ., Park Ridge, NJ.

Darrington, H. 1987. A long running cereal. Food Manuf. 62(3):47.

Davies, D. 1986. Sorting out quality—In colour. Frozen Chilled Foods 40(11):32.

Decelles, G. A., and Larson, V. M. 1973. Process for preparing clustered, mixed ready to eat cereal products. U.S. patent 3,868,471.

DeLauder, W. R., and Spring, F. E. 1976. Process for preparing a protein fortified natural cereal. U.S. patent 4,097,613.

dos Santos, C. 1979. Method and apparatus for producing a masa product. U.S. patent 4,221,340.

Dougherty, M. E., Jr. 1988. Tocopherols as food antioxidants. Cereal Foods World 33:222.

duBois, M. 1988. Natural flavor ingredients. Cereal Foods World 33:560.

Edwards, L. W. 1977. Co-crystallization of dextrose and sucrose on cereal products. U.S. patent 4,101,680.

Englyst, H. N., Anderson, V., and Cummings, J. H. 1983. Starch and non-starch polysaccharides (NSP) in some cereal foods. J. Sci. Food Agric. 34(12):1434.

Fast, R. B. 1967. Process for preparing a coated ready-to-eat cereal product. U.S. patent 3,381,706.

Fast, R. B. 1987. Breakfast cereals: Processed grains for human consumption. Cereal Foods World 32:241.

Fast, R. B., Hreschak, B., and Spotts, C. E. 1971. Preparation of ready-to-eat cereal characterized by honey graham flavor. U.S. patent. 3,554,763.

Ferrara, P. J. 1977. Method of modifying the properties of cereal flours and the modified flours so produced. U.S. patent 4,145,225.

Fragas, R. R. 1977. Corn powder preparation. U.S. patent 4,199,612.

Garbutt, J. T. 1978. Isolation of proteinaceous materials. U.S. patent 4,163,010.

Gaubert, R. J. 1977. Product wrapping machine and method. U.S. patent 4,148,170.

Giguere, J. R. 1978. Method of degerminating a kernel of grain by simultaneously compressing the edges of the kernel. U.S. patent 4,189,503.

Gilbertson, D. 1976. Sweet coatings for food products. U.S. patent 4,089,984.

Gobble, H. G., Vandell, R. M., and Mooi, R. 1979. Method of making a ready-to-eat breakfast cereal. U.S. patent 4,178,392.

Godshall, M. A. 1988. The multiple roles of carbohydrates in food flavor systems. Cereal Foods World 33:913.

Gormley, T. R. 1985. Oats—Human food par excellence. Farm Food Res. 16(3):89.

Gralak, B. G. 1975. Process for preparing protein supplemented flavored instant grits. U.S. patent 3,974,295.

Greger, J. L. 1988. Calcium bioavailability. Cereal Foods World 33:796.

Haag, R. A., Rousseau, P. M., and Martin, T. O. 1965. Cereal process and product. U.S. patent 4,075,356.

Haas, G. H. 1973. Ready-to-eat cereal containing debittered soy products. U.S. patent 3,920,852.

Haines, P. S., and Popkin, B. M. 1987. Cereal products in the school lunch program. Cereal Foods World 32:197.

Hart, E. R. 1978. Method of processing corn. U.S. patent 4,234,614.

Hauck, B. W., and Huber, G. R. 1989. Single screw vs twin screw extrusion. Cereal Foods World 34:930.

Hayward, J. R., Keyser, W. L., and Zielinski, W. J. 1978. Cereal protein fortified food bar. U.S. patent 4,145,448.

Henthorn, L. J., and Kincs, F. R. 1974. Breakfast cereals. U.S. patent 3,814,822.

Hirzel, R. W., Olmstead, A. W., and Howard, W. C. 1975. Method and apparatus for producing lapped shredded food articles. U.S. patent 4,004,035.

Hoppner, K., and Lampi, B. 1983. The biotin content of breakfast cereals. Nutr. Rep. Int. 28(4):793.

Hoppner, K., and Verdier, P. 1984. Contribution of breakfast cereals to the daily intake of folacin, pantothenic acid and biotin. Can. Inst. Food Sci. Technol. J. 17(2):121.

Hoseney, R. C. 1986. Principles of Cereal Science and Technology. Am. Assoc. Cereal Chem., St. Paul, MN.

Hughes, J. H. 1978. Continuous production of starch hydrolysates. U.S. patent 4,221,609.

Hus, J. Y. 1977. Preparation of pasta. U.S. patent 4,208,439.

Ishida, J. 1978. Method of producing instant cupped noodles. U.S. patent 4,166,139.

Jimenez, J. A. 1978. Apparatus for preparing taco shells. U.S. patent 4,184,418.

Jimenez, J. A. 1979. Method of cooking foods. U.S. patent 4,189,504.

Johannson, H. P. 1978. Powdered gluten composition and process for the production thereof. U.S. patent 4,150,016.

Johnson, L., Gordon, H. T., and Borenstein, B. 1988. Vitamin and mineral fortification of breakfast cereals. Cereal Foods World 33:278.

Jones, G. P., Briggs, D. R., and Toet, H. 1983. The mono- and disaccharide content of some breakfast cereals. Food Technol. Aust. 35(6):281.

Kelly, V. J. 1978. Stabilization of rice polish. U.S. patent 4,158,066.

Kelly, V. J., and Smalligan, W. 1973. Process for preparing instant cereals and the resulting product. U.S. patent 3,887,714.

Kelly, V. J., Thompson, P. J., and Smalligan, W. J. 1977. Barley malt sterilization. U.S. patent 4,140,802.

Kent, N. L. 1983. Technology of Cereals. Pergamon Press Ltd., Oxford, UK.

Kickle, H. L., Ball, W. J., Jr., and VonSchanefelt, R. 1977. Processed vegetable seed fiber for food products. U.S. patent 4,181,747.

Klopfenstein, C. F. 1988. The role of cereal beta-glucans in nutrition and health. Cereal Foods World 33:865.

Kremperheide, O. F. 1977. Method and apparatus for husking and drying cereal and legume kernels. U.S. patent 4,196,224.

Ladbrooke, B. D., Quick, G. R., and Singer, N. S. 1978. Wheat-based lipoprotein complexes and methods of making and using same. U.S. patent 4,200,569.

LaGrange, V., and Sanders, S. W. 1988. Honey in cereal-based new food products. Cereal Foods World 33:833.

Lane, R. P. 1985. The application of flavors to extruded products. Perfum. Flavor. 10(4):53.

Lawrence, N. F., and Reesman, S. H. 1975. Corrugated cereal flakes. U.S. patent 3,998,978.

Levine, L. 1988. Understanding extruder performance. Cereal Foods World 33:963.

Liepa, A. L. 1972. Soy flour breakfast cereal. U.S. patent 3,687,687.

Linzberger, R., Ketting, L., and Maechler, E. 1977. Apparatus for the grinding of cereal. U.S. patent 4,140,285.

Longenecker, J. G. 1977. Apparatus for shaping and precooking of tortillas. U.S. patent 4,241,648.

Lorenz, K. 1984. Cereal and dental caries. Page 83 in: Advances in Cereal Science and Technology, Vol. 6. Y. Pomeranz, ed. Am. Assoc. Cereal Chemists, St. Paul, MN.

Lu, S. 1987. Sorghum based breakfast cereals. Ph.D. thesis, Univ. of Nebraska, Lincoln. (Diss. Abstr. Int. B47(7):2699)

Lu, S., and Walker, C. E. 1988. Laboratory preparation of ready-to-eat breakfast flakes from grain sorghum flour. Cereal Chem. 65:377.

Luallen, T. E. 1988. Structure, characteristics, and uses of some typical carbohydrate food ingredients. Cereal Foods World 33:924.

Lusas, E. W., and Rhee, K. C. 1986. Applications of vegetable food proteins in traditional foods. ACS Symp. Ser. 312:32.

Lutz, R. J. 1973. Extruded cereal product and its preparation. U.S. patent 4,044,159.

Lyall, A. A., and Johnston, R. J. 1974. Emulsified oil and sugar cereal coating and incorporating same. U.S. patent 3,959,498.

Martin, T. O. 1975. Process for preparing a dried agglomerated cereal mixture. U.S. patent 4,038,427.

Martin, T. O., and Clausi, A. S. 1980. Rice cereal and process. U.S. patent 4,238,514.

Matz, S. A. 1959. Chemistry and Technology of Cereals as Food and Feed. AVI Publ. Co. Inc., Westport, CT.

Matz, S. A. 1970. Cereal Technology. AVI Publ. Co., Inc., Westport, CT.

Maxwell, D. 1988. The impact of ingredients upon extruder performance. Prep. Foods 157:177.

McAuley, J. A., Hoover, J. L. B., Kunkel, M. E., and Acton, J. C. 1987. Relative protein efficiency ratios for wheat-based breakfast cereals. J. Food Sci. 52:1111.

McAuley, J. A., Kunkel, M. E., and Acton, J. C. 1987. Relationships of available lysine to lignin, color and protein digestibility of selected what-based breakfast cereals. J. Food Sci. 52:1580.

McDermott, L. P. 1988. Near infrared reflectance analysis of processed foods. Cereal Foods World 33:498.

McFarlane, I. 1988. In-line measurement and closed-loop control of the color of breakfast cereals. Cereal Foods World 33:978.

Mendoza, F. C. 1977. Apparatus for molding and precooking corn and wheat tortillas. U.S. patent 4,197,792.

Meredith, F. I., and Caster, W. O. 1984. Amino acid content in selected breakfast cereals. J. Food Sci. 49:1624.

Midden, T. M. 1989. Twin screw extrusion of corn flakes. Cereal Foods World 34:941.

Miller, B. D. F. 1988. Drying as a unit operation in the processing of ready-to-eat breakfast cereals: I. Basic principles. Cereal Foods World 33:267.

Miller, B. D. F. 1988. Drying as a unit opration in the processing of breakfast cereals: II. Selecting a dryer. Cereal Foods World 33:274.

Miller, R. C. 1988. Continuous cooking of breakfast cereals. Cereal Foods World 33:284.

Monahan, E. J. 1988. Packaging of ready-to-eat breakfast cereals. Cereal Foods World 33:215.

Mongeau, R., and Brassard, R. 1984. Effects of dietary fiber from shredded and puffed wheat breakfast cereals on intestinal function in rats. J. Food Sci. 49:507.

Morgan, K. J. 1978. An hedonic index for breakfast cereals. Ph.D. thesis; University of Missouri-Columbia, Columbia, MO. (Diss. Abstr. Int. B)

Morgan, K. J., and Goungetas, B. P. 1988. Cereal food usage: Impact of socioeconomic and demographic characteristics. Cereal Foods World 32:433.

Muller, R. 1977. Method of degerming maize. U.S. patent 4,229,486.

Mulvaney, S. J., and Hsieh, F.-H. 1988. Process control for extrusion processing. Cereal Foods World 33:971.

Nadeau, D. B., and Clydesdale, F. M. 1986. Effect of acid pretreatment on the stability of citric and malic acid complexes with various iron sources in a wheat flake cereal. J. Food Biochem. 10(4):241.

Niemann, E.-G., Christoffers, D., and Georgi, B. 1977. Method and apparatus for determining the total protein content or individual amino acids. U.S. patent 4,135,816.

Niewoehner, C. 1988. Calcium and osteoporosis. Cereal Foods World 33:784.

Olson, R. D., and Eifler, R. H. 1975. Breakfast cereal process and product. U.S. patent 3,976,793.

Orr, D. 1984. The role of twin extrusion cooking in the production of breakfast cereals. Inst. Chem. Eng. Symp. Ser. 84:61.

Oughton, R. W. 1978. Process for the treatment of comminuted oats. U.S. patent 4,211,801.

Oughton, R. W. 1979. Treatment of comminuted oats under the influence of an electric field. U.S. patent 4,208,259.

Oughton, R. W. 1979. Treatment of comminuted proteinaceous material under the influence of an electric field. U.S. patent 2,408,260.

Paugh, G. W. 1973. Continuous puffing method. U.S. patent 3,908,034.

Pellaton, R. C. 1977. Pasta noodle packaging apparatus and method. U.S. patent 4,147,081.

Peterson, R. D. 1984. Survey results of consumer understanding and usage of nutrition labels on ready-to-eat breakfast cereal in Canada. J. Can. Diet. Assoc. 45:344.

Pierce, C. W. 1986. Micronized wafer. U.S. patent 4,153,733.

Pitchon, E. 1980. Food extrusion process. U.S. patent 4,225,630.

Powell, H. B. 1956. The Original Has This Signature—W. K. Kellogg. Prentice Hall, Englewood Cliffs, NJ.

Rao, G. V., and Shoup, F. K. 1978. Method for fractionating the whole wheat kernel by sequential milling. U.S. patent 4,201,708.

Reesman, S. H. 1975. Corn flake process and product. U.S. patent 4,013,802.

Reesman, S. H., and Lawrence, N. F. 1975. Cereal flakes product and process. U.S. patent 3,996,384.

Rosenquist, A. H., Knipper, A. J., and Wood, R. W. 1975. Method of producing expanded cereal products of improved texture. U.S. patent 3,927,222.

Sakakibara, S., Sugisawa, K., Kimura, T, Yasukawa, T., and Kamo, T. 1979. Method for preparing instant cooking noodles. U.S. patent 4,234,617.

Satake, T. 1976. Rice pearling apparatus with humidifier. U.S. patent 4,133,257.

Satake, T. 1977. Rice pearling machine with humidifier. U.S. patent 4,148,251.

Satake, T. 1978. Rice pearling apparatus. U.S. patent 4,155,295.

Schade, H. R., Baggerly, P. A., and Woods, D. R. 1975. Frosted coating for sweetened foods. U.S. patent 4,079,151.

Scharschmidt, R. K., and Murphy, L. 1977. Flake cereal process and product. U.S. patent 4,211,800.

Schwab, E. C., Petersen, W.D., and Bumbiers, E. 1973. Process for making high-protein cereal. U.S. patent 3,873,748.

Schwengers, D., Bos, C., and Andersen, E. 1978. Method of preparing refined starch hydrolysates from starch-containing cereals. U.S. patent 4,154,623.

Seiler, D. A. L. 1988. Microbiological problems associated with cereal based foods. Food Sci. Technol. Today 2(1):37.

Seligsohn, M. 1983. The industry in 1983: What's ahead? Food Eng. 55(2):22.

Shanbhag, S. P., and Szczesniak, A. S. 1981. Fruit and cereal products and process therefor. U.S. patent 4,256,772.

Shemer, M. 1979. Novel physical form of gluten, method for its manufacturing and its uses. U.S. patent 4,238,515.

Shin, S. H. 1976. Process for preparing steam-kneaded vermicelli products. U.S. patent 4,181,746.

Sloan, A. E. 1988. Change in breakfast pattern may be among current consumer trends. Cereal Foods World 32:246.

Smith, K. T. 1988. Calcium and trace mineral interactions. Cereal Foods World 33:776.

Specker, B. L., and Tsang, R. C. 1988. Vitamin D in infancy. Cereal Foods World 33:788.

Stoll, W. F. 1978. Lasagna noodle. U.S. patent 4,166,136.

Suggs, J. L., and Buck, D. F. 1978. Pasta conditioner. U.S. patent 4,229,488.

Swientek, R. J. 1987. Extrusion laboratories. Food Process. 48(1):90.

Tims, O. L. 1979. Flour tortillas. U.S. patent 4,241,106.

Toma, R. B., and Curtis, D. J. 1989. Ready-to-eat cereals: Role in a balanced diet. Cereal Foods World 34:387.

Tomadini, G. 1976. Transport system with swinging or vibrating belts for the drying of long pasta. U.S. patent 4,167,068.

Torres, A. 1978. Flour substitutes. U.S. patent 4,219,580.

Tressler, D. K. 1970. Oat breakfast cereal. U.S. patent 3,494,769.

Velasco, R. E., Jr. 1978. Method and apparatus for producing masa corn. U.S. patent 4,205,601.

Verberne, P., Zwitserloot, W. R. M., and Nauta, R. R. 1977. Method for the separation of wheat gluten and wheat starch. U.S. patent 4,132,566.

Vincent, M. W. 1983. Manufacture of biscuits & confectionery products by extrusion cooking. Confect. Prod. 49(2):107.

Visser, J., and Bams, G. W. P. 1980. Protein fibres. British patent 1,574,448.

Vollink, W. L., Kenyon, R. E., Barnett, S., and Bowden, H. J. 1979. Freeze dried cereal fruit. Canadian patent 845,642

Walker, C. E. 1989. The use of sucrose in ready-to-eat breakfast cereals. Cereal Foods World 34:399.

Walon, R. G. P. 1978. Process for separating and recovering vital wheat gluten from wheat flour and the like. U.S. patent 4,217,414.

Weaver, C. M. 1988. Calcium and hypertension. Cereal Foods World 33:792.

Wenlock, R. W., Sivell, L. M., and Agater, I. B. 1985. Dietary fibre fractions in cereal and cereal-containing products in Britain. J. Sci. Food Agric. 36(2):113.

White, E. G. 1978. Process for manufacturing a whole wheat food product. U.S. patent 4,179,527.

White, J. S., and Parke, D. W. 1989. Fructose adds variety to breakfast. Cereal Foods World 34:392.

Wiedmann, W., and Strobel, E. 1988. Extrusion cooking compared with conventional food processes. Food Mark. Technol. June:34.

Witt, P. R., Jr. 1978. Torrefied barley for brewer's mashes. U.S. patent 4,165,388.

Wolffing, R. M., Batten, C. J., and Harris, M. C., Jr. 1977. Farina milling process. U.S. patent 4,133,899.

Yoshida, K., Hatanaka, Y., Kudo, K., and Aoki, T. 1978. Process for producing quick-cooking noodles. U.S. patent 4,230,735.

Zabik, M. E. 1987. Impact of ready-to-eat cereal consumption on nutrient intake. Cereal Foods World 32:234.

Index